Research Collaboration

A step-by-step guide to success

Research Collaboration

A step-by-step guide to success

Annette Bramley

Liz Ogilvie

A how-to guide for collaborative organizations, teams and individuals

IOP Publishing, Bristol, UK

© IOP Publishing Ltd 2021

All rights reserved. No part of this publication may be reproduced, stored in a retrieval system or transmitted in any form or by any means, electronic, mechanical, photocopying, recording or otherwise, without the prior permission of the publisher, or as expressly permitted by law or under terms agreed with the appropriate rights organization. Multiple copying is permitted in accordance with the terms of licences issued by the Copyright Licensing Agency, the Copyright Clearance Centre and other reproduction rights organizations.

Permission to make use of IOP Publishing content other than as set out above may be sought at permissions@ioppublishing.org.

Annette Bramley and Liz Ogilvie have asserted their right to be identified as the authors of this work in accordance with sections 77 and 78 of the Copyright, Designs and Patents Act 1988.

ISBN 978-0-7503-2727-5 (ebook)
ISBN 978-0-7503-2725-1 (print)
ISBN 978-0-7503-2728-2 (myPrint)
ISBN 978-0-7503-2726-8 (mobi)

DOI 10.1088/978-0-7503-2727-5

Version: 20211201

IOP ebooks

British Library Cataloguing-in-Publication Data: A catalogue record for this book is available from the British Library.

Published by IOP Publishing, wholly owned by The Institute of Physics, London

IOP Publishing, Temple Circus, Temple Way, Bristol, BS1 6HG, UK

US Office: IOP Publishing, Inc., 190 North Independence Mall West, Suite 601, Philadelphia, PA 19106, USA

Contents

Preface		vii
Acknowledgements		ix
Author biographies		x
1	**The 'what' and 'why' of research collaboration**	**1-1**
1.1	Why collaborate in research?	1-1
1.2	UCL Ventura: providing continuous positive airway pressure devices for Covid-19	1-2
1.3	What is collaboration?	1-5
1.4	Collaborations and relationships	1-6
1.5	Collaboration and co-operation	1-9
1.6	Challenges of co-produced collaborative research	1-14
1.7	Key learning points—chapter 1	1-17
	References	1-18
2	**Get ready: preparing yourself to collaborate**	**2-1**
2.1	Mindsets	2-1
2.2	'I should be so lucky'—the role of serendipity and how you can encourage it	2-10
2.3	The key communication skills that make collaboration	2-13
2.4	Listening and its role in collaboration	2-20
2.5	Key learning points—chapter 2	2-26
	References	2-27
3	**Leading by example: preparing your team to collaborate**	**3-1**
3.1	Diversity and collaboration	3-2
3.2	The role of leaders in research collaboration	3-4
3.3	Behaviours of people with good leadership skills	3-6
3.4	Leading researcher or collaborative leader?	3-10
3.5	Power dynamics and their impact on collaboration	3-14
3.6	Leaders and followers in collaborations	3-16
3.7	The i-sense-AHRI collaboration and the importance of leadership	3-17

3.8	The 'soft stuff'—psychological safety and why it's important for collaboration	3-18
3.9	The importance of 'belonging'	3-21
3.10	How to go about creating a psychologically safe atmosphere?	3-21
3.11	Key learning points—chapter 3	3-24
	References	3-25

4   Creating a collaborative organisation   4-1

4.1	Organisations as complex systems	4-2
4.2	Creating a collaborative organisation using the cultural web model	4-5
4.3	Using the cultural web as an analytical tool	4-11
4.4	Leading a collaborative organisation	4-11
4.5	Properties of systems that inhibit collaboration	4-15
4.6	Putting the cultural web into practice—real world approaches to creating a collaborative organisation	4-17
4.7	Thinking about collaboration, the cultural web and collaborating with industry	4-23
4.8	Some thoughts about the research and innovation system of systems in the UK	4-25
4.9	Key learning points—chapter 4	4-29
	References	4-30

5   What's next for research collaboration?   5-1

5.1	Impact of Covid-19 and climate change on how we collaborate	5-1
5.2	Remote collaboration	5-3
5.3	Increasing inclusion and flattening hierarchies	5-4
5.4	Planning remote workshops	5-6
5.5	Engineering serendipity into online collaboration	5-9
5.6	Future directions for research collaboration?	5-10
5.7	Key learning points—chapter 5	5-16
	References	5-17

Preface

We all collaborate with others all of the time; whether it's with our families, our friends, or at work.

'The urge to form partnerships—to link up in collaborative arrangements—is perhaps the oldest, strongest and most fundamental force in nature. There are no solitary free living creatures: every form of life is dependent on other forms.' [1]

Although we all collaborate with others, that doesn't mean it is easy—it isn't! Most family gatherings can be sources of joy, stress, frustration, and support all at the same time. It's the same with research collaborations. They can be exasperating, overwhelming, and difficult. But more importantly they can be rewarding, inspiring, and transformational.

Our aim is to give you more confidence and understanding about entering into research collaboration so that you too can begin to embark on seriously exciting research endeavours. We want to help you understand the building blocks of a successful collaboration. We want to help you go on to lead, build, and be part of successful and effective research collaborations and organisations, that will go on to make a real difference to the global challenges we face.

As the amount of available knowledge and information increases, collaboration will allow researchers to be even more agile and flexible in their approach to problem solving; bringing together people, their skills, and their expertise in unique ways. This should be an efficient and effective way of delivering projects, yet time and time again we see that collaborations have struggled for one reason or another. The difficulties that we encounter when we try to collaborate mean that at times our efforts stall—a phenomenon known as collaborative 'inertia' [2].

Being able to collaborate is a foundational and transferable practice that drives organisations to be better at problem solving, more innovative, less siloed, and more effective. Collaboration is not only the engine that powers team research, it can also help create cultures that are more inclusive, curious, and connected. Yet it is precisely because teamworking is so ubiquitous in workplaces that collaborative practice often gets taken for granted.

To combat this, universities can teach collaborative practice to their students and researchers as part of continuing their professional development; keep abreast of the latest understanding of meta-research in research collaboration and adopt enabling technologies, skills, and infrastructures. We can all learn from the growing body of work about what enables collaboration to be successful, both in business and in research. We can use the good practice that does exist to inform our own practice, and that of our teams, co-workers, and students. In this way we can build an upward skills spiral where everyone is better able to collaborate, and these capabilities can be shared between all kinds of disciplines and organisations.

In this book we explore some of these issues in a research context and suggest some practical steps that you can take to help people bring their unique contributions to your collaboration. We hope that reading our book will enable you to make a distinctive contribution to a collective endeavour. We want to give you some tools and techniques that will reduce the uncertainty about stepping forward into new relationships and give you the confidence to experiment, learning from what doesn't turn out quite how you expected!

We believe that successful collaboration is an ongoing practice, full of learning. This book is called *Research Collaboration: a step-by-step guide to success*, for two reasons. Firstly, we wanted it to be a useful, practical reference, that you can use to support your own practice. Secondly, we believe that collaboration works best if you take it literally, step by step, starting small and aiming high. Building relationships, trust, and common values. Making mistakes, learning from them, and trying again. And repeat.

We hope that you find helpful insights and tools in this book that you can use both now and in the future.

Annette Bramley and Liz Ogilvie, October 2021.

References

[1] Thomas L 1980 *Key Report.* **46** 1–3
[2] Huxham C and Vangen S 2004 *Organ. Dyn.* **33** 190–201

Acknowledgements

We would like to thank everyone that we have interviewed for this book, allowed us to reproduce their figures, provided jokes, and those that have taken the time and trouble to provide feedback. In no particular order this includes: Rebecca Shipley, Tim Baker, Mervyn Singer, John Girkin, Helen Szoor-McIlhenny, Julie Sanders, Stuart Humphries, Adrian Mulholland, Matthew Avison, Annella Seddon, Robert Hughes, Simeon Yates, David Petley, Tony Ryan, Chris Hewson, Mike Trenell, Ruth Patchett, Rachel McKendry, Michael Thomas, Eleanor Gray, Valerian Turbé, Isabel Bennett, Ben Miller, David Payne, Sebastien Ourselin, Louise Heathwaite, David Sweeney, Anne Moore, Robert Thompson, Hilary Noone, Lorna Mackay, Anne-Marie Coriat, Claire Spreadbury, Katy Turner, Jim Spencer, Ian Craddock, Nick Goldspink, Richard Plenty and Terri Morrissey.

We would like to thank Joe McEntee, IOP Publishing's Consultant Editor for commissioning this work and for helpful encouragement along the way.

Liz would like to thank Caragh Dewis, Scott Middleton and Bob Ogilvie. Annette would like to thank Dan Eastell, Angus Eastell, Susie Eastell, Anne Farrow, Rachel Woolley, Caroline Batchelor, Joe de Sousa, Jon Cooper and Kev Dhaliwal.

We would both also like to thank all the people and organisations we have worked with over the years who have inspired our passion for multidisciplinary research collaborations, from whom we have learned so much and with whom we have made so many memories. There are too many of you to name individually—we are humbled by and grateful for all your support. Thank you.

Author biographies

Annette Bramley

At the helm of the N8 Research Partnership since 2018, Annette is its Director and Chief Collaboration Officer. She is a public speaker highly regarded for her expertise in research culture and collaboration.

Annette is an ambassador for excellent collaborative research that has a genuine impact in the world. Her particular passion is multidisciplinary partnership, team science, and collaboration.

With a first class degree and DPhil in Materials Science from the University of Oxford, she brings real world understanding from the laboratory combined with the experience of more than 20 years guiding researchers from a range of disciplines at the Engineering and Physical Sciences Research Council (now part of UK Research and Innovation). Her LinkedIn recommendations (https://bit.ly/3lUKPK6) give a sense of how her input has helped make a huge range of people and projects successful.

Annette is also an artist in the mediums of embroidery and acrylic whose work has been exhibited in London, Cornwall, and the Isle of Wight. She holds the Certificate in Technical Hand Embroidery from the Royal School of Needlework—awarded with Distinction. She is currently studying for a Professional Diploma in Group Sound Therapy with the British Academy of Sound Therapy.

Liz Ogilvie

Liz has over 20 years' experience in facilitation and leadership development. Liz is one of the Founding Directors of The Collective Facilitation Ltd. She has worked for global multinationals such as Procter and Gamble, education institutions including Kingston College, and micro-enterprises in under-developed parts of London. Her career has seen her facilitate sandpits throughout the world, helping to inspire innovative, transformative multidisciplinary solutions that answer increasingly complex challenges. She has worked with organisations in the UK including UK Research and Innovation, Wellcome, the Royal Society, the British Academy, Cancer Research UK, Cystic Fibrosis Trust, and Autistica. Globally her clients have included the World Health Organisation and the National Science Foundation of the United States of America.

She is a school governor and Chair of the Governors of a Community Health Organisation.

IOP Publishing

Research Collaboration
A step-by-step guide to success
Annette Bramley and Liz Ogilvie

Chapter 1

The 'what' and 'why' of research collaboration

In this section we focus on the context and reasons for collaboration—the 'why?' of successful collaborations. We will investigate what productive collaboration might look like and the key factors that make collaborations work. We will explore the nature of collaborative relationships, the importance of co-production and co-creation, and the differences between collaboration and co-operation. We will introduce some concepts here without going into too much detail and develop them later in the book.

1.1 Why collaborate in research?

The large, global challenges of our time are complex and systemic. No single researcher, no single discipline, no single organisation has the answer. We must work together.

The concept of 'collaborative advantage' is 'the synergy that can be achieved by combining the unique resources and expertise of one organisation with that of others' [1]. We can replace the word organisation in this definition with researcher. We then derive a definition of research collaborative advantage as 'the synergy that can be achieved by combining the unique resources and expertise of one researcher with that of others'.

Simeon Yates, Professor of Digital Culture and Associate Pro-Vice-Chancellor for Research Environment and Postgraduate Research at the University of Liverpool, told us about one of his favourite illustrations of the importance of multidisciplinary collaborative research; understanding the impact of the Apple iPhone.

'If you really want to understand the impact the iPhone has had on our society, well, you need to understand the engineering. You need to understand why it's rectangular. Why aren't iPhones round or oval? What does production think? And what are people doing with them? Well, you need social scientists. How do people interact with them? Well, you need a psychologist. What are the long-term health

effects? Well, you need health professionals and so on. How do they impact learning? We need educationalists. You can't answer the question about the impact of the iPhone from one place or discipline. That's why it's important to work across disciplines.'

The successful development of a vaccine for Covid-19 depended on the ability of researchers to collaborate with others, as an article from the *New York Times* in April 2020 eloquently points out:

'While political leaders have locked their borders, scientists have been shattering theirs, creating a global collaboration unlike any other in history. Never before have so many experts in so many countries focused simultaneously on a single topic.' [2]

One of the silver linings of the Covid-19 pandemic has been seeing the walls and barriers that get in the way of research collaboration crumble before our very eyes. The story of University College London (UCL) Ventura is the story of just one of many collaborations that emerged in response to the pandemic. It captures the spirit of researchers and their ability to form relationships in the middle of a crisis, driven by shared goals and values.

1.2 UCL Ventura: providing continuous positive airway pressure devices for Covid-19

In a conference call on Monday 16 March 2020, the UK Government launched its ventilator challenge. The aim, to secure 30 000 ventilators for the NHS to treat Covid-19 patients, within a fortnight. Ministers were said to be desperate [3]; as they believed that tens of thousands of people across the United Kingdom might require ventilation. Rebecca Shipley, Professor of Healthcare Engineering and Head of the Institute for Healthcare Engineering at UCL, was invited to help with this mammoth effort.

Nowadays, a ventilator is a very sophisticated piece of medical technology. It requires trained operatives and dedicated bedspace for sedated patients in intensive care units. Mervyn Singer, Professor of Intensive Care Medicine at UCL and an Intensive Care Consultant with the University College Hospitals NHS Trust (UCH), could see a flaw in this plan. Even if the Government could secure the additional ventilators, it would be difficult to make use of them because of the staffing and beds that would also be needed.

At this point in time, there was relatively little experience in treating Covid-19 in the UK, but in parts of Italy and China this new infection had been spreading rapidly. In China, a new 1000-bed hospital was built from scratch in about 10 days to handle the number of patients that they were treating in Wuhan province alone. In the Lombardy region of Italy, intensive care units (ICUs) were treating around ten times more patients than usual [4]. Doctors in China and Italy were desperately warning their counterparts in other countries about the sheer numbers of people that would need hospital treatment because of the virus. They had valuable insights into treating Covid-19 gained from intense weeks of front-line practice in hospitals in their own countries. They were keen to pass on their knowledge, in the hope that more lives could be saved.

Mervyn had friends and acquaintances within these hospitals and knew that they had been quickly overwhelmed by the demand for and lack of supply of ventilators. He knew that more ventilators would be needed. Importantly, though, he learned something else from his colleagues on the front-line in Italy and China. He heard how they had used non-invasive respirators to help support patients' breathing, increase the oxygen content in their bloodstream and keep them out of the ICUs.

This was how Mervyn came to understand that by using what are known as continuous positive airway pressure (CPAP) devices, patients with breathing difficulties due to Covid-19 could be treated with a blend of air and oxygen to their lungs. CPAP could be delivered without invasive ventilation and sedation, and without taking up valuable beds in intensive care, which could be reserved for the most unwell patients. Mervyn heard, although there was no hard data available, that colleagues in Italy were convinced that these devices had been lifesavers because they had protected the critical care capacity and saved it for the most unwell.

The only trouble was that CPAP devices, although available in the UK, were in very short supply at the start of the pandemic. Indeed, in March 2020 UCH had only 12 stand-alone CPAP devices. Mervyn knew that they would need many, many, more.

Rebecca Shipley and a colleague, Tim Baker, Team Principal at UCL Racing, knew Mervyn and wanted to get his expert ICU experience for the ventilator challenge. They met late on a Friday to catch up, and to the engineers' surprise, Mervyn threw down a gauntlet, or more literally, a CPAP device. He told them, 'I don't want a ventilator, I'd rather have a CPAP device, something that will be quick, simple and easy to use and easy to manufacture'.

What's more, Mervyn was able to provide an example of something he'd used in the 1990s, a very simple mechanical CPAP device. This 'whisperflow' device was now off-patent and could be reverse engineered and analysed. By so doing, new devices could be designed, built and taken through the UK's stringent regulatory approvals process at pace.

The UCL engineers needed no further bidding. Within 24 hours, they had pulled together a team which included Mercedes–AMG High Performance Powertrain (a Formula 1 motorsport team), UCL postgraduates, staff and even some alumni who returned to support the effort.

They worked around the clock to create the first prototype in just 100 hours and obtained regulatory approval of the device in a remarkable 10 days. As Tim told us, 'we had already delivered a prototype and obtained MHRA approval when other groups were still at the stage of signing non-disclosure agreements'.

You can read more about this amazing achievement in their paper published in *The Lancet* [5]. The team also went on to produce a Mk II device that re-engineered the patient circuits for a more comfortable patient experience, and significantly reduced oxygen utilisation for the hospitals, conserving a valuable resource.

This collaboration was heroic. Not only are the UCL Ventura CPAPs being used in over 150 NHS hospitals in the UK, the consortium also released full details of the designs and manufacturing instructions at no cost to governments, industry manufacturers, academics, and health experts across the globe. Over 1950 teams in 105 countries have downloaded the design, saving lives and creating new skills and manufacturing operations [6, 7] worldwide thanks to UCL Ventura. One of their collaborators, Niels Birbaumer from Smart City Tech in Paraguay, summed it up when he said,

'What UCL has achieved in the UK and around the world is what humanity is all about—helping each other without expecting anything in return. Together, we can create a real impact across the globe.'

Let's look at what made this collaboration work so well and so quickly:
1. *A clear focus and goal*: Mervyn's understanding of the benefits of a CPAP device compared to a ventilator, and his ability to locate and deliver to his collaborators an example of a simple device gave the healthcare engineering team a clear focus for what needed to be achieved. Rebecca told us 'the motivations for all of the partners were so clear and so unifying, it was not hard to persuade people to get involved. People were desperate to have a way to contribute'.
2. *Relationships*: The partnerships between UCH and UCL were already in place; through UCL's Institute of Healthcare Engineering, led by Rebecca Shipley. Mervyn had relationships with colleagues in China and Italy while Tim had a long-term network of existing contacts in the manufacturing sector, particularly with a former student who had since become Head of Mechanical Engineering at Mercedes–AMG High Performance Powertrain. Over the years Tim had made a point of engaging industry colleagues with course development and student assessment, bringing them into the physical university space. Tim told us, 'the relationships had been there for many years so it was literally just a phone call [to begin this collaboration]'. Rebecca said, 'retrospectively it felt lucky that all the relationships were in place; often in these scenarios you have to build the team and build the relationships, but this team was *good to go*'. Was it luck? We don't think so. We think that what the UCL Ventura team was experiencing was a return on investment in their relationships, and this is something we will return to later in the book.
3. *Trust*: Mervyn, Rebecca, and Tim already knew and trusted each other and were happy to get together informally to share openly views about what was needed, even when these didn't conform to the vision of the Ventilator Challenge.
4. *An ability to play to partners' strengths*: Mercedes brought the mentality of a race team to the project, driving it forward at pace, together with the facilities to rapidly design and prototype on a commercial production line. When the device went into full production, 40 machines that would normally produce F1 pistons and turbochargers were used for production of the CPAP

devices, and the entire Mercedes Brixworth facility was repurposed to meet this demand.
5. *Adding value to every individual partner*: Genuine research collaboration like UCL Ventura leads to outputs and impacts that couldn't have been achieved by the collaborators working on their own. The collaboration enriched and added value, above and beyond the intellectual input of the partners and it inspired massive commitment. Without doubt everyone involved in this project was professionally and personally changed by the experience. Andy Cowell, Managing Director of Mercedes–AMG High Performance Powertrains, said: 'It is exceptionally pleasing to see that the flow devices swiftly engineered and produced in volume here at Brixworth are helping patients around the UK. The supply of devices to the local Northampton hospital engendered a great sense of pride for the whole team.'

Hopefully, this story will have inspired you as much as it inspires us. We will explore the key building blocks for collaboration further and in more detail in other sections of this book.

First though, we need to take a step back and build a shared understanding of the concept of collaboration and the terms, or jargon, that we will be using as we move through the rest of the book.

1.3 What is collaboration?

Collaboration is, by definition, something you can't do on your own.

The word 'collaboration' is derived from the Latin word 'collaborare'—literally 'to work together'—and is usually associated with an intellectual endeavour like research, or a creative practice such as art, drama, or writing [8]. This book is itself an example of the outputs of a collaboration between two friends and colleagues—although its focus is another type of collaborative activity—the practice of research collaboration.

Every person, collaboration, and project is unique. Some collaborations come to fruition quickly and some are a slower burn. Some never come to fruition at all because the conditions for one or more of the parties weren't quite right. Every attempt at collaborating is an opportunity to learn and to grow, both personally and professionally. It is an investment in your future, in your development and your own perspectives as a researcher and a human being. It's making a deposit in the bank of social capital.

In the 2019 Wellcome Report on research culture, collaboration was identified as one of nine characteristics that are part of a good research culture [9]. While collaboration and openness were often the best indicators of a healthy working culture, in more toxic environments collaboration was often the first characteristic to be lost. Our hope is, that by putting in place the conditions and cultures needed for research collaboration, we can also put in place conditions for a healthier, more diverse research culture.

1.4 Collaborations and relationships

Research collaborations, like families and friendships, are built on relationships, relationships that have the power to create change, break down silos and to challenge institutions, structures, systems, and prejudices. Powerful stuff. No wonder that sometimes getting involved in collaborations can seem daunting and it's much easier to put them on the 'too difficult' pile.

Jeremy Roschelle and Stephanie Teasley, education researchers in the United States, noted that:

'Collaboration does not just happen because individuals are co-present; individuals must make a conscious, continued effort to coordinate their language and activity with respect to shared knowledge.' [10]

Because most research collaborations involve working with people and relationships, this means that there are always human dynamics to manage and negotiate. Embarking on collaborative research adventures means that you must start thinking in an emotionally intelligent as well as intellectually intelligent way. Logic and rational argument are key tools for an academic researcher; and to collaborate effectively academic researchers also need to develop emotional intelligence with key skills of empathy, intellectual tolerance and humility. These are not skills or values which are widely taught or rewarded through traditional recognition and promotion frameworks. Indeed, these skills may be seen as 'soft', perhaps more associated with the feminine side of our personalities and have suffered as a result in the dopamine-driven, journal impact factor chasing, world of academic research.

At one workshop Liz was facilitating, a participant came up to her and said, 'I am not sure anyone will want to talk to me—they have all done so much and have so much to give—why would anyone talk or want to work with me—as compared to everyone else I have so little to give?'

This is a concern shared by many early career researchers who can often feel intimidated by more established researchers whose reputations can come into the room along with them.

We have found that when the right enabling environment is created, it is amazing how often early career researchers are absolutely central to novel and, perhaps, unexpected collaborations that look at challenges through a new 'lens'.

In collaboration it is important to hold true to a belief that everyone has something to give and contribute. It is important to look beyond status, to do away with hierarchies that you bring with you, and not to create new ones.

If we don't create the right conditions, it can be all too easy for dominant personalities and extrovert characters to take over. Later in the book we will explore how you can create the dynamics and interactions we need for healthy collaboration.

1.4.1 Teamwork and collaboration

Does being part of a team mean that *de-facto* you are part of a collaboration? The UK Civil Service College says no, there is a difference between teamwork and collaboration. They set out an excellent description of the difference between the two [11]:

'Both teamwork and collaboration involve a group of people working together to complete a shared goal.'

'The key difference between the collaboration and teamwork is that whilst teamwork combines the individual efforts of all team members to achieve a goal, people working collaboratively complete a project collectively. Those collaborating work together as equals, usually without a leader, to come up with ideas or make decisions together to complete a goal. Whereas teamwork is usually overseen by a team leader, and those within a team are delegated individual tasks to complete to contribute towards the team's end goal.'

We would contend that leadership is crucial for successful research collaboration. Nonetheless, the essence of the definition here—bringing a group of people to *come up with ideas together* to complete a goal—is what research collaboration is about. In their article, the UK Civil Service College also highlights the importance of interpersonal skills for collaboration, because:

'...it's very important that every person involved contributes their ideas, opinions, and knowledge.'

Whenever we work in a team with other people to achieve something, we need to invest time and effort in the team itself before we can all start pulling together. In management circles this process is often known as the team lifecycle of, 'forming, storming, norming, and performing'. This requires an investment of effort, time, and other resources, without an immediate payback. The return on the investment, of course, comes in the longer term as teams become 'high performance teams' [12].

In today's ever-changing world most teams do not get the opportunity to move beyond the forming phase, before their members are moved into new teams and the process of forming starts again. Teams are often brought together for a short period of time to focus on an issue or a challenge. Once the problem is solved, the team is disbanded. This can sometimes be very successful, if there is a task or a problem that requires quite linear thinking, or to solve problems which are very similar, or analogous, to ones that people in that team have solved before. Things get trickier when the problems are complex, ill-defined or 'wicked' problems, but more of that in a little while.

An 'all in it together' mentality in a team can cause quite strong functional bonds to form quickly. Annette experienced this first-hand when working with colleagues across the UK Research Councils, Innovate UK, and UK Office for Life Sciences to come up with proposals for the second wave of the UK Industrial Strategy Challenge Fund. Working together intensely over a few weeks, with a known end goal, colleagues with highly developed skills came up with coherent propositions very quickly. There were benefits to working together in this way; the team developed new connections in their networks, many enduring past the particular project. The team members also learned a lot about the contemporaneous priorities of other organisations in a short space of time. They pulled together to produce what was asked of them but did not have the time or space to really get to know each other. What had been learned about the priorities of others would, inevitably, become out of date with the passage of time.

In research organisations, people have to participate in many different teams. You probably recognise the feeling of having lots of different priorities competing for your time and attention. It can be hard to carve out the time you and collaborators need together to create relationships, clarity of purpose, or meaning when you are under pressure on many fronts. The Nobel Prize winning economist Daniel Kahneman explored the impact that stress and pressure can have on our logical thinking in his book, *Thinking Fast and Slow* [13]. Many of you will have come across his concepts in training related to unconscious bias.

When we are under pressure, our brains tend to use fast, highly myelinated neural connections which have been laid down and reinforced through previous experience; rather than go to the more effortful, slow process of using weaker or building new neural connections. What this means for innovation is that teams which are not able to have the time and space to truly collaborate are unlikely to be as innovative as they have the potential to be. There are situations when this is absolutely fine, when 'good enough' is, aptly, good enough.

Where we start to experience problems is when we ask teams to deliver innovative or transformational proposals but do not provide them with the time and space to move from teamwork to collaboration. The total delivered is unlikely to be more than the sum of the individual parts and those who have sponsored the collaboration may come away with the feeling that the collaboration has under-performed. The teams rarely make the transformational impact that they could have if they had been true collaborations.

Teaching 'team-skills' has often been used as a proxy for teaching collaboration. Some of the skills required to be a good team leader, or a good team player may also help with research collaboration, but on their own they are not sufficient, particularly if they have only been developed at a relatively superficial level on a 'team-building course'.

We believe that collaboration is an ongoing practice that takes time and repeated engagement between people. When we recognise collaboration as a practice this means that it is not simply a single skill that can be taught, that there is no unique formula for success and certainly no 'golden bullet'. It does mean that each collaboration is an opportunity to learn and build on a variety of skills, and to navigate each project relationship by relationship. We will expand on the crucial role of leadership, listening, and mindsets for collaboration at a team and organisational level in chapters 2, 3 and 4.

A key component of interpersonal skills and emotional intelligence is self-awareness. As you start to build research collaborations with others, you can usefully start to build self-awareness, an understanding of your own preferred way of working and your own responses to the problem that the collaboration will try and address. We will also give you some tools for this in subsequent chapters.

1.5 Collaboration and co-operation

The dictionary definition of co-operation is 'the fact of doing something together or of working together toward a shared aim'. This is remarkably similar to the definition of collaboration. So are collaboration and co-operation the same thing?

In an article for the *Huffington Post* [14], former advertising executive and entrepreneur Lynn Power talks about co-operation being the division of work amongst participants so that each individual is responsible for solving a portion of the problem. Kip Kelly, an executive coach in the United States writing with the rock-star turned entrepreneur Alan Schaefer in a white paper for the University of North Carolina, agrees [15]. They explain that co-operation is when each person on a team develops their own plans and shares those plans with the team. There may be a joint discussion, but the focus remains on individual actions and achievement rather than a collective strategy.

This is essentially what happens in conventional team-working, so team-members co-operate to achieve an overall outcome. We often see this type of team-working in research programmes. The grant is won and the money is divided for each contributor to do 'their bit'. The project can be very successful, delivered to time and on budget and achieve its goals. The whole is intended to be, and is realised as, 'the sum of its parts'.

Kip Kelly and Alan Schaefer define collaboration as when individual goals are subordinated for collective achievement. Lynn Power defines collaboration as

'the co-ordinated synchronous activity that is a result of a continued attempt to construct and maintain a shared conception of a problem.'

But what does this really mean, and how does it translate from advertising to research?

What Lynn Power is saying is that collaboration involves an iterative approach to problem-solving. In her view, this way of working is one where, as people work on a problem together, they bring feedback to try and understand the issue, and take account of the learnings to reframe and update the understanding of the whole team. In turn this might result in a different investigative path to the one that was originally anticipated. This sounds very much like a typical collaborative research project and is where the real 'added value' of the collective effort can be realised.

In research the distinction between co-operation and collaboration is often less marked than in the examples quoted by Lynn Power. Research programmes involving more than one person probably lie on a spectrum from wholly co-operative to wholly collaborative. Many researchers would consider themselves to be collaborating when they are working together with at least another person, as per the dictionary definition. Part of the reason that we tie ourselves in knots when it comes to understanding research collaborations is that there are so many potential variations and contexts—from the particle physicists collaborating on large, international machines to the poet and computer scientist working together to look at the concept of intelligence.

Lynn Power also notes that there are further factors which need to be in play for her version of true collaboration:

'True collaboration is hard—and it doesn't mean compromise or consensus-building. It means giving up control to other people. It means being vulnerable. It means needing to know when to fall on your sword and when to back down. Collaboration is inherently messy.'

Giving up control, being vulnerable and backing-down are behaviours that are not rewarded in academic culture. Individuals need to be able to 'let go', which some people will find easier than others. You have to forget territoriality or 'hunkering down in bunkers' and move into a new joined and shared territory.

This may go part of the way to explaining why some types of research collaboration can be hard to get off the ground. Sometimes you may have to unlearn behaviours that have been taught over many years and adopt new ones. There can be uncertainty about how other colleagues, not involved in the collaboration, will react to your ideas and methods as a result.

The collaborative method, a more iterative or 'design-led' approach to research, may also be more challenging for some parts of the research base than others. This type of methodology is used frequently in the arts and engineering but less congruent to the 'hypothesis-driven' methodologies of the physical and biomedical sciences. A typical worry is how to frame a research programme or proposal in the collaborative context and convincing different research communities, simultaneously, of the appropriateness of the approach. How peer review views an approach which challenges the conventional or established ways of thinking and doing, and in turn impacts whether a grant is awarded or a paper is published, is part of optimising a collaborative system, something we will pick up in one of the later chapters.

1.5.1 Collaboration and co-opetition

Co-opetition is a particular form of collaboration between competing organisations or businesses [16]. Literally a mix of co-operation and competition, this form of collaboration is a type of strategic alliance that changes the rules of competition to move from a zero-sum game to a plus-sum game. The theory behind co-opetition is based in game theory, statistical models that consider the ways that synergy can be created by partnering with competitors, thus creating an environment where the end results mean that all partners benefit.

The N8 Research Partnership is an example of co-opetitive practice between eight research intensive universities in the North of England. Co-opetition is particularly common in the tech sector, but has been adopted by companies as diverse as Google and Yahoo, Ford and General Motors, DHL and UPS. A particularly recent example of co-opetition is in the development of the Pfizer-BioNTech vaccine for

Covid-19. The agreement to partner between the two, normally competing, businesses has enabled the manufacturing capacity for this novel vaccine to be dramatically increased.

The key to successful co-opetition is understanding the drivers and variables that influence whether the rival partners are more likely to compete or to collaborate. It requires that there is some partial alignment of interests, for example:

- Sharing costs of research and development, particularly at the pre-competitive stage. A good example of this is UK Water Industry Research (UKWIR) [17]—the collaborative research platform for the water companies in the UK and Ireland.
- Access to complementary skills, capabilities or equipment.
- Avoiding duplication of effort.
- Risk sharing.
- Growing the overall market for a product or service by setting standards or interoperability protocols.
- Addressing cultural or employment practices in a sector.

As with all collaborations, there are challenges and risks, which are heightened where the co-opetitors are also in direct competition for funding, students, or customers. Trust, knowledge and data sharing and protection, governance and power are all issues that need to be overcome.

Collaboration with rivals also has an emotional aspect. The positive side is that progress in tackling the most difficult issues can be made more quickly. The shadow side is that some people can find it difficult to collaborate with those that they have previously considered adversaries. Indeed, one definition of collaboration is 'traitorous co-operation with an enemy' [18]. Some people simply find the concept of the plus-sum game, added value or win–win difficult to grasp. As with all types of collaboration, partners need to adopt the right mindsets if the co-opetition is to be successful. We'll come back to mindsets and how to prepare yourself, and others, to collaborate in the next chapter.

Intermediary organisations, like N8 Ltd, the core team at the heart of the N8 Research Partnership, or UKWIR can help the competing partner organisations overcome the challenges of collaboration by acting as neutral platforms, or honest brokers. The intermediary organisation can define what sorts of information will be shared and protected, create the conditions for trust between the partners and help facilitate mutually beneficial outcomes. They can help identify people with the right mindsets and create the environment needed for win–win opportunities to be identified and nurtured. Adam Brandenburger and Barry Nalebuff, the authors of the seminal work on co-opetition [19], note in their 2021 article for the *Harvard Business Review*:

*'Ultimately getting the right mindset means choosing the right people...need to staff the co-opeting teams with people open to the dual mindset of co-opetition. That isn't always easy, because people tend to think in either/or terms, as in either compete or cooperate, rather than compete **AND** cooperate. Doing both at once*

requires mental flexibility, it doesn't come naturally. But if you develop that flexibility and give the risks and rewards careful consideration, you may well gain an edge over those stuck thinking only about competition.'

1.5.2 Collaboration and co-production

Co-production and co-creation are both forms of collaboration between people with diverse skills and lived experiences.

For the purposes of this book we're using this definition of co-production:

'A collaborative, iterative process in which those affected by the research are active partners in conducting it; working together problems are defined and ideas are shared and improved;'

More succinctly:

'Nothing about us, without us', or a focus on

'what matters to *us, not what the matter is* with *us'*

The latter two definitions are often used in biomedical and health-related research as well as in social science research as part of patient and public involvement in research—a type of co-production process.

More and more funders outside of these areas are coming to recognise the benefits of co-production done ethically and genuinely. It is driven by the need for funders of research to demonstrate to their stakeholders, whether government, fundraisers, donors or the taxpayer, that research and innovation is an essential investment. There can also be financial benefits; for example, engaging business users of research can lead to private sector leverage on government funds while simultaneously increasing the proportion of a country's gross domestic product that is invested in research and innovation.

Mary Stuart, who recently retired as Vice Chancellor of the University of Lincoln, said in her vision for the 21st Century Lab [20]:

'If we are to stay relevant we need to anticipate and prepare for change and work with our communities to shape and drive the 21st Century as it continues to unfold. Knowledge is no longer created in ivory towers but it is shared and developed in multiple partnerships. Universities need to engage with these different centres of knowledge. Their melting pot of research, knowledge exchange and teaching activities puts them in a unique position to provide the thinking, talent and workforce that can ride the wave of change. Not only that but, crucially, it can inform its direction to create positive outcomes for our world.'

In co-production the problems and the communities of interest themselves define the research agenda. They are, absolutely, experts through experience.

Asking harder and better, more relevant questions can generate new fundamental insights which would not have otherwise been possible. The process of co-production is explicitly inclusive, bringing in partners from diverse backgrounds. There are many more opportunities to engage people from marginalised communities and to conduct research in a socially responsive way. This exchange of knowledge and skills can engage communities more with their local universities, lead to cross-cultural learning, more open attitudes and in turn improve lives in the short and longer term.

Alongside this, co-production can enable access to information and data which would not otherwise be available to academic researchers. In the UK, the National Institute for Health Research (NIHR) Surgical Med Tech Co-operative [21] has had patient involvement as a fundamental part of its programme from the start. Patients were part of the steering groups deciding the research agenda, as well as commenting on the research methodology and resulting medical technologies.

We saw co-production play a key part in the EPSRC Interdisciplinary Research Collaboration 'A Sensor Platform for Healthcare in the Residential Environment (SPHERE)' [22], led by Ian Craddock, Professor of Data Driven Health at the University of Bristol, with partners at the Universities of Reading and Southampton. SPHERE focused on creating a sensor platform for the home to meet the challenge of increasing numbers of people with chronic co-morbid health conditions which require continuous management outside of a hospital setting. This highly successful collaboration brought together clinicians, engineers, designers, and social care professionals as well as members of the public (residents of Bristol). Engaging with people who would eventually support the project by consenting to have a platform installed in their homes would help the research team understand what might be acceptable in terms of data collection, analysis, and storage. It also helped to refine the design of a wearable sensor device for service users with dementia, with features that were not always found in mass market devices. Long battery life, for example, was very important to this group as they did not want to have to remember to charge their devices or where they had put the charger. These were concerns that sometimes did not resonate with some academics who were used to regularly charging their own laptops, phones, and smartwatches, but were real life issues for the people who need these technologies most.

As well as helping to design better research programmes, co-production is about involving the communities, people, or sectors that will benefit from research to ensure that they benefit in a way that makes a tangible difference to their lives. Co-producing knowledge gives a clearer purpose to research and innovation, and a better understanding of how research outcomes might be used in the real world. It leads to better, more relevant, and faster impacts. It helps to mitigate against unforeseen behaviours and outcomes, like the prosthetic leg that lives under the bed because it is too uncomfortable to wear.

Co-production has had most traction in health research over recent years, but it is just as relevant in other parts of the research and innovation landscape.

In the UK, the N8 Policing Research Partnership [23] co-produced research between the Universities of Liverpool and Lancaster working with Merseyside Police and the charity Women's Aid. This has led to a better understanding of coercive control and a new learning tool for the police to help officers identify this form of domestic violence. Police are better able to support victims as a result.

Again in the UK, as part of the N8 AgriFood Programme [24], researchers from the Universities of Sheffield and Lancaster worked with food hubs across the North of England. They figured out the best way to set up food hubs, looking at what works for different cultures and places, and how to deal with the stigma around food poverty. Together the partners ran food hubs, cookery classes, and local food cafes in local communities, collecting evidence on what worked and developed free action packs for people wanting to set up new food hubs elsewhere.

Co-production also enables the research agenda to evolve throughout the course of a research and innovation programme in response to the insights of all the partners. This not only means that the outcomes are more likely to be adopted and used, it also means that the research can be much more responsive to users in fast-changing and uncertain environments, where the requirements may change very quickly.

In 2019, Liz facilitated a workshop on behalf of the Wellcome Trust and the World Health Organisation Global Taskforce on Cholera Prevention. Funded research teams were brought together with representatives from the communities who are at risk from, and prone to, cholera outbreaks every year, and from the governments of the countries where these people lived. The aim was to ensure that the research would be relevant and would make a difference to the lives of people who live with the threats and consequences of cholera.

A single workshop and the understanding that was generated as a result enabled early changes to be made to the research programmes which in the long term would make them more impactful. Research questions were better defined, time and resources were not wasted pursuing avenues that were unlikely to address the real problems, and ultimately the research would impact positively on health outcomes thus saving lives and money. Partnerships and collaborations were formed during the workshop that would enable co-production to happen throughout the research programme, not just at the outset, so that new knowledge generated through the programme could be understood and its implications worked through together.

1.6 Challenges of co-produced collaborative research

So why isn't everyone carrying out co-produced collaborative research?

Well, of course, co-production isn't suitable for every research project or collaboration, so it is very important to select the appropriate methodology—and the right partners—for the research project. There are complexities—many of which can be addressed by adopting some of the behaviours and skills that you will read about in this book.

Without doubt co-production is a fundamentally different type of research, which can challenge accepted practices and definitions of 'excellence'. It means that ALL partners, and the types of knowledge and experience that they each bring, have to be respected and valued equally. It means that time and resource have to be invested in building relationships, establishing a common purpose, managing power differentials, building a common language, and establishing trust.

No longer are the research findings the sole output of the research programme; they are just one form of knowledge and impact that emerges from co-produced research which goes on to make a difference to the people that matter most. And, of course, all of this needs to be balanced with rigour. Collaborative and co-produced research must show both rigour and relevance; a bigger hurdle than for discovery research for which rigour alone is sufficient.

Perhaps some of the biggest challenges to co-production—and possibly the most difficult to tackle—are the cultural norms that pervade some academic disciplines and organisations. These can lead to unconscious biases and in turn impact on the broader innovation ecosystem. Of course, people in communities, industry, and government also have their own biases about universities and academic researchers, perhaps seeing them as being 'out of touch', or 'not relevant' or speaking a language that they can't relate to or understand. Building relationships and overcoming these stereotypes takes time and resource, and the timelines for return on these investments can be long.

Co-producing research is, without a doubt, more time-consuming than discovery research. Expressed more positively, it takes time and a lot of investment and work from the outset to build relationships, understanding and to develop a common purpose. However, community, public and industry engagement can no longer be seen as an add-on at the end of a project. As we saw in SPHERE, true and authentic co-production is a two-way equal partnership and the benefits of this are felt by everyone involved.

All of this leads to challenges for universities, funders, publishers, and peer review. How best to assess co-produced research and how to recognise the contributions and intellectual input of everybody involved? Could these criteria and processes themselves be co-produced through working with partners in the broader research ecosystem? How can we create organisations where collaboration is part of, 'the way we do things' and where co-production is part of the collaborative practice? We will come back to these thorny systemic issues later when we take a look at the collaborative organisation.

1.6.1 Is collaboration transactional or transformational?

All collaborations need clear ground rules and transparent processes so that everyone knows how they will work together and how decisions will be made. But overly process-driven collaborations can kill creativity and wipe out added value by creating hierarchies and bureaucracy. The transactional processes need to become habit, part of the culture or 'the way we do things around here', so that they don't get in the way and that the transformation can happen.

Collaborations are more likely to be transformational when participants can let go of the need to control and be controlled. The transformational collaborations that we have seen become successful enable risk-taking, learning from what doesn't work and have an honest approach to knowledge, with it being OK not to know the answers to everything. Whether or not it is safe to let go of control and to be honest affects whether you can collaborate as an individual, as a team and within or across organisations. You will see this strand developed in every section of the book.

1.6.2 And finally…

We leave you with one final but essential belief—don't strive for the perfect collaboration—it doesn't exist! Relationships are not an end point, and life is not a rehearsal. Perfection is the enemy of the good and learning happens in the mistakes we make along the way. By experimenting and trying things out, we build trust and connection with our collaborators, we reduce the pressure on ourselves and others and importantly, we can enjoy the experience.

1.7 Key learning points—chapter 1

- Collaboration is an ongoing practice, built on relationships that takes time and repeated engagement between people.

- To make a collaboration work you need:
  - a clear focus and goal,
  - relationships,
  - trust,
  - to play to partners' strengths,
  - to add value to every individual partner.

- Emotional and intellectual intelligence are key skills for research collaboration.

- Creating the right enabling environment for early career researchers is important and can lead to novel insights by looking at challenges through a different 'lens'.

- Collaboration shares some features of teamwork; but team skills are not sufficient for collaboration.

- Moving from team-working to collaboration takes time.

- True collaboration means sharing power and control with others.

- Intermediary organisations can help competing groups to collaborate by acting as honest brokers.

- Co-producing research with communities of interest has many benefits but is not easy. It requires that all types of knowledge and experience are valued equally. It takes time and resource to build these collaborations, which have to demonstrate both rigour and relevance.

References

[1] Huxham C and Vangen S 2004 *Organ. Dyn.* **33** 190–201 (Elsevier)
[2] Apuzzo M and Kirkpatrick D 2020 Covid 19 Changed how the world does science together *New York Times* April 14
[3] https://theguardian.com/business/2020/may/04/the-inside-story-of-the-uks-nhs-coronavirus-ventilator-challenge
[4] https://thelancet.com/journals/lancet/article/PIIS0140-6736(20)32154-1/fulltext
[5] Singer M, Shipley R, Baker T, Cowell A, Brealey D and Lomas D 2020 *Lancet Respir. Med.* **8** 1076–78 (Elsevier)
[6] https://ucl.ac.uk/healthcare-engineering/news/2020/nov/teams-paraguay-progress-rapidly-local-cpap-manufacture
[7] https://innovationaction.org/cpap-in-africa/
[8] *Oxford English Dictionary* (Oxford: Oxford University Press)
[9] Moran H and Wild L 2019 Wellcome Research Culture Qualitative Report @ShiftLearning
[10] Roschelle J and Teasley S D 1995 Construction of shared knowledge in collaborative problem solving *Computer-Supported Collaborative Learning* ed C O'Malley (New York: Springer) pp 69–97
[11] https://civilservicecollege.org.uk/news-understanding-the-differences-between-teamwork-and-collaboration-203
[12] Google 2015 The five keys to a successful Google team re:Work (November) Re:work.withgoogle.com
[13] Kahneman D 2011 *Thinking, Fast and Slow* (New York: Farrar, Strauss and Giroux)
[14] https://huffpost.com/entry/collaboration-vs-cooperat_b_10324418
[15] http://execdev.kenan-flagler.unc.edu/hubfs/White%20Papers/unc-white-paper-creating-a-collaborative-organizational-culture.pdf
[16] Brandenburger A and Nalebuff B 2021 The rules of co-opetition *Harvard Business Review* **99** 00178012
[17] https://ukwir.org/leading-the-water-industry-research-agenda
[18] https://lexico.com/definition/collaboration
[19] Nalebuff B J and Brandenburger A M 1997 *Co-Opetition* (London: Profile Business)
[20] *Thinking Ahead Exploring the Challenges and Opportunities of the 21st Century*, University of Lincoln https://lincoln.ac.uk/home/media/responsive2017/documents/thinking-ahead-university-of-lincoln.pdf
[21] https://surgicalmic.nihr.ac.uk/
[22] https://irc-sphere.ac.uk/
[23] https://n8prp.org.uk/
[24] https://n8agrifood.ac.uk/

IOP Publishing

Research Collaboration
A step-by-step guide to success
Annette Bramley and Liz Ogilvie

Chapter 2

Get ready: preparing yourself to collaborate

'A big benefit of open interaction is that it inspires collaboration, whether it simply involves a reagent, an approach, or even a way of thinking that does not come naturally to you but enhances the impact of what can be learned'—Professor Rebecca Heald, University of California, Berkeley [1]

In this chapter we will look at various factors that can help you, as an individual, collaborate more successfully in research.

Every collaboration is about bringing together a group of individuals around a common focus and goal. Each person brings their whole self into the collaboration—their attitudes, beliefs, values, experiences (both good and bad), mindsets, and aspirations. We'll start to unpack what you might choose to take with you into a collaboration and give you some ideas about what you might like to leave behind.

There will also be an external context, or system, that you're working in. You'll bring with you your perspective and understanding of this context. Sometimes we call this being able to see the 'bigger picture' or we talk about having a particular 'world view'—the way someone sees the context through the lens of their position and experiences.

Somehow in collaborations we manage to make sense of all this and build a common way forward ... with a dash of serendipity for good measure (figure 2.1).

In this section we will consider:
- the importance of mindsets;
- the key communication skills of collaboration, language, listening, and asking the right questions;
- the role of serendipity.

2.1 Mindsets

The first thing we need to do is to ensure that you approach any collaboration with the right mindsets. Here we will look at proactive and defensive mindsets, scarcity

Figure 2.1. You bring your mindset, values, network, and knowledge into a collaboration—which sits within a bigger context.

and abundance mindsets, the growth mindset and the importance of curiosity. We will reflect on their impact on research collaboration and how you can get yourself into the 'right' mindset to collaborate.

2.1.1 Proactive and defensive mindsets

If you aspire to be more successful in research collaboration it will help to actually be looking to become involved in collaboration. You might think that this goes without saying, but we have both facilitated consortium-building workshops where some of the attendees have either been 'sent' by someone more senior than them, or been driven by a sense of obligation, or even a feeling that they don't want to miss out on something.

This kind of mindset is a **defensive mindset**. Someone approaching a collaboration with a defensive mindset doesn't really want to be in collaboration, although they may want the perceived rewards for being a successful collaborator. They don't come with an open mind, ready to learn and to experiment the way forward. Instead, their mindset acts as a protective shield which 'blocks' new information and 'protects' their thinking and approach. In their book, *Uncertainty Rules*, Richard Plenty and Terri Morrissey visualise the defensive mindset [2] (figure 2.2).

People that come to multidisciplinary workshops with a defensive mindset show up ready to defend their world view, and their thinking can be very 'black and white'. They can find it very difficult to cope with the ambiguity and 'shades of grey' in the conversations. Other views and perspectives may be interpreted as a personal attack on their integrity and way of thinking. You can recognise defensive mindset behaviours when people repeatedly use language like: 'the reason why we can't do this is ...', or, 'we've tried that before and it didn't work'.

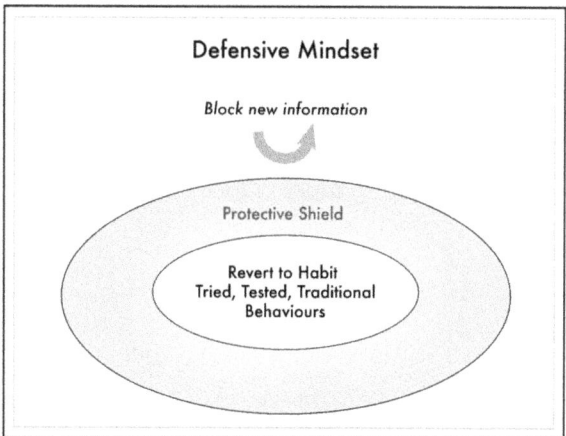

Figure 2.2. A defensive mindset. With permission from Richard Plenty and Terri Morrisey [2], copyright Cork University Press.

Stuart Humphries, Professor of Evolutionary Biophysics at the University of Lincoln, summed this up well: 'There's an academic mindset where if you don't agree with something you feel that you have to argue your case. I often don't feel that need, and so I think that helps me work with people.'

Someone with a defensive mindset will probably be viewed by other participants and potential collaborators as a 'blocker'. Being unwilling to experiment, flex or look for opportunities to bring perspectives together is indicative of someone trying to cling on to the status quo and not really wanting to let go. A defensive mindset rarely results in collaboration or genuine innovation.

You may be able to think of times when you have gone into a situation with a defensive mindset. We all sit on different points on a spectrum and can be fiercely loyal to some ideas in some situations. We are more likely to adopt a defensive mindset if we are put in a situation we don't really want to be in, or where we perceive there to be a threat to our safety. It doesn't make us a bad person, but it will make entering into collaboration really, really difficult. When our protective shield is in place, we can have hard time empathising with others on the other side because their behaviour just doesn't make sense to us.

The opposite of a defensive mindset is a **proactive mindset**. If we can shift our thinking and behaviours to those of a proactive mindset, we are much more likely to be successful in our attempts to collaborate.

In Richard Plenty and Terri Morrisey's visualisation of the proactive mindset, the defensive shield has been dropped and instead there is a flexible style grounded in a strong core character, ethics, and values (figure 2.3).

It will not surprise you that a proactive mindset is a much more empowering mindset for collaboration. People with this type of mindset often use language like, 'what if we did …?', 'that's a great idea and would be even better if …', 'that's interesting', 'that sounds exciting', 'how could we work together to make a difference?' The language uses high energy words which bring a sense of moving

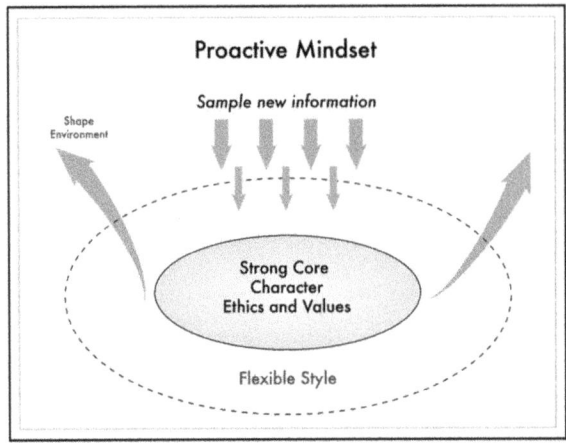

Figure 2.3. A proactive mindset. With permission from Richard Plenty and Terri Morrisey [2], copyright Cork University Press.

forward together, with people building on each other's ideas to create something new and different.

People with proactive mindsets are more likely to be able to identify opportunities to try out their ideas within the 'shades of grey' of what is known and not known. They are more comfortable dealing with ambiguity and uncertainty that exists as they feel their way forward in an iterative approach to the collaboration, increasing their commitment as they go. It's not an 'all-or-nothing' approach. In the previous chapter we spoke about how the collaborative method is a more design-led approach to research which may be more familiar to some parts of the research base than others. No matter what your disciplinary background, by adopting a proactive mindset you can flex your style and approach to enable you to experiment with working with others.

Another key behaviour of someone with a proactive mindset is that they understand the 'bigger picture', or context, of a problem and are open to other peoples' perspectives in that context. They may be able to imagine themselves in another person's shoes and able to empathise with their experiences.

So how can you try and make sure that you, and others, come into opportunities for collaboration with a proactive, rather than a defensive, mindset?

Preparation is part of the secret sauce. Choosing a new and exciting area to work in, rather than well-trodden ground, will help people to be able to leave their protective shields behind and will feel less threatening. It can also be a bit of a leveller in terms of status and experience, which can help contribute to psychological safety, which we talk more about in the next chapter. An important challenge or problem to solve can have a similar effect.

If you are embarking on a new collaboration or attending a workshop on a specific challenge, you might like to consider carrying out your own PESTLE analysis to help you prepare. A PESTLE analysis (also known as a STEEPLe analysis) looks at the Political, Economic, Social, Technological, Legal and

Environmental factors that might influence the collaboration or the challenge. You can find lots of free templates to download online which give you a framework to record your thinking. This will help you get a feel for the 'big picture' of the environment in which an issue, or potential collaboration, is situated. This exercise may help you start to see some alternative perspectives and can help identify potential areas of focus.

You can also use a PESTLE analysis as the basis for discussions at a workshop to help unpick the influences of external factors, and parts of the system, on your collaboration. It helps to pinpoint where opportunities might exist for possible collaborations and the barriers you might need to overcome.

2.1.2 Scarcity and abundance mindsets

Steven Covey, the author of *7 Habits of Highly Effective People*, wrote,

'*Most people are deeply scripted in what I call the Scarcity Mentality. They see life as having only so much, as though there were only one pie out there. And if someone were to get a big piece of the pie, it would mean less for everybody else.* **The Scarcity Mentality is the zero-sum paradigm of life.**'

'*People with a Scarcity Mentality have a very difficult time sharing recognition and credit, power, or profit—even with those who help in the production. They also* **have a very hard time being genuinely happy for the successes of other people**— *even, and sometimes especially, members of their own family or close friends and associates. It's almost as if something is being taken from them when someone else receives special recognition or windfall gain or has remarkable success or achievement.*'

There are many causes of a scarcity mindset. Fear, competition for resources and uncertainty trigger the part of our brain that is in charge of our flight or fight response. When something is scarce, our brain thinks it must be valuable. Cue desperate stockpiling of toilet roll in the face of the Covid-19 pandemic, for example, or people not sharing their knowledge or expertise. These are Darwinian urges—we don't want to be left behind. Fear of missing out, or FoMo, is a relatively new variant of scarcity—'am I missing out on something?'.

This kind of mindset has important consequences, and they tend not to be positive, so it is not a surprise that it has a highly destructive effect on our ability to collaborate and share. A scarcity mindset leads to lapses in self-control, less willpower, over-consumption, poor judgement, and envy. It gives us tunnel vision—an obsessive focus on the thing that is scarce and which we desire. It drives short-term over long-term thinking, often making matters worse instead of better in the long run. And worst of all for collaboration, it makes us selfish and greedy. We keep those scarce resources for ourselves or for our tribes. It is hard to be generous if you feel insecure.

We have seen this first-hand in workshops when participants talk about an idea that they have but are very protective of it and are reluctant to share with the group. They hang on in the belief that if they share their idea they will lose control, or that someone will steal it. They worry that they could lose the intellectual property, or the grant funding as a result. The paradox is that sharing an idea often results in it growing, developing and ultimately improving, thanks to questions, challenge, and new perspectives.

Academia is a breeding ground for a scarcity mindset. There is competition for funding, to be published in high impact journals, for status, for prizes and recognition. Imposter Syndrome—or the fear of being found out—can trigger bad behaviours and in some cases can prevent people putting themselves forward for collaborative opportunities. Unconscious biases can cause people to focus on success, take action too quickly, try too hard to fit in and depend too much on 'experts' [3]. Demand management on behalf of funders can elicit reactions like, 'what's best for me/my group/my university?', when what we need for collaboration is, 'what is the best research idea that we can come up with that will make the biggest impact?'.

There are benefits to real scarcity. It can drive innovation and collaboration, as we have seen play out during the recent pandemic as academics and industry have come together to devise new ventilation devices and diagnostics for Covid-19. The need for innovation and the common cause can break down barriers which at other times may have seemed insurmountable and drive new creative approaches.

The irony is that thriving during real scarcity requires what Stephen Covey calls 'an abundance mindset'—to appreciate what we have and make the most of it, being generous with our time, knowledge, and resources, sharing with others and thinking about life as an adventure. Brené Brown, a research professor at the University of Houston, says that we need to be careful to realise that the opposite of scarcity is not 'more than you could ever imagine, but enough' [4].

Katia Verresen, a leadership coach, has summarised the difference between the scarcity and abundance mindsets [5] (figure 2.4).

We have seen through Covid-19 that with common goals and an abundance mindset we can move mountains. We need to take these lessons and change our behaviours and systems to make collaboration abundant within our research and innovation ecosystem in the UK.

2.1.3 Curiosity

'The mind that opens to new ideas never returns to its original size.'—Albert Einstein

'If a man will begin with certainties he shall end in doubts; but if he
will be content to begin with doubts, he shall end in certainties.'—Francis Bacon

'Curiosity encourages new ways of thinking, challenges long-held assumptions
and fuels transformative change.'—Francesca Gino, Rebel Talent

Most successful collaborations happen because of curiosity—it's the coming together of people who ask interesting questions and are willing to take the measured risk of engaging. In the case studies we have researched for this book we have seen this over and over again. So many of the successful collaborations have

	Scarcity	Abundance
Point of View	You're a victim, a bully, or simply checked out.	You're in the driver's seat.
Physical Energy	Contracted body, tense shoulders, clenched jaw, short of breath.	Relaxed and alert, expansive posture. Rooted and balanced, present, breathing deeply and evenly.
Emotional Energy	Draining energy in the room and in your interactions. Feeling frustrated, impatient, anxious, afraid, angry, overwhelmed and powerless. Giving power over to group-think and pressure.	Feeling empowered, engaged, positive, like you're working on something bigger than yourself. You energise and inspire others. You're excited about the challenges and growth ahead.
Mental Energy	Confused, disorganised, narrow in your thinking, only focusing on what's not working. Typical thought pattern: "I have no choice".	Feeling of clarity, the ability to perceive multiple angles, listen actively, and notice what others are not seeing. Flexible and adaptable. Typical thought pattern: "I always have a choice. If I were to notice something new, what would it be?" Creative agency. Non-judgemental beginner's mind.

Figure 2.4. The characteristics of scarcity and abundance mindsets adapted from [5].

started by researchers, stakeholders and end users being curious as a collective, having conversations and asking questions. They found common ground from their different backgrounds and ultimately this led to research which led to new knowledge and interesting, high quality outputs.

Francesca Gino, in her book *Rebel Talent* [6], explains the performance benefits of curiosity in all sorts of jobs, and illustrates this with examples from diverse sectors of business, including the curiosity of Harry Houdini! She has found that curious people often end up being star performers for several reasons: they have larger networks, they are comfortable asking questions and they more easily create and nurture ties with others. Gino found in her research that these ties were critical for the career development and success of the star performers, and as we'll see later, ties are really important for nurturing serendipity too.

Curiosity is a key driver of creativity and innovation and a key part of a growth mindset—a tendency to be open to learning from challenge and failure. The growth mindset was introduced by Stanford University psychologist Carol Dweck in her seminal work, *Mindset—the new psychology of success* [7]. More recently, researchers have used continuous electroencephalography (EEG) to image the neural activity in people with fixed and growth mindsets when they performed tasks and made mistakes. They found much less brain activity in people with fixed mindsets, whereas those with growth mindsets actively process errors [8] (figure 2.5).

Much has been written on the subject of growth mindsets, including their importance for academic success [9]. Here we will focus specifically on curiosity because of its importance for collaboration, and direct interested readers to the many books and articles about the growth mindset that are already available.

Adam Steltzner is the NASA engineer who led the programme that developed the Mars rover, appropriately named 'Curiosity', that landed on the so-called 'red'

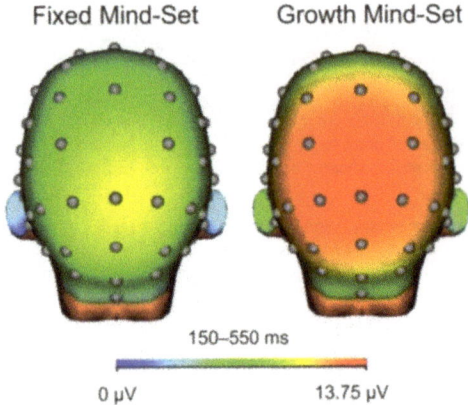

Figure 2.5. EEG voltage potential maps illustrating the difference in neural activity between fixed and growth mindsets reprinted from Moser *et al* [8] copyright SAGE Publications.

planet in 2012. He's currently the Chief Engineer for the NASA Mars2020 mission and the rover 'Perseverance', that landed on the red planet in February 2021. He says that 'a curious mind stays agile, innovative and competitive'.

Who does not want to go into a research collaboration being agile, innovative and competitive? Curiosity can lead to new and unpredictable insights, help break habitual thinking, challenges assumptions and helps us to notice more patterns in our environment. Children are naturally curious, and curiosity can be viewed as a playful, creative mindset. It is associated with a sense of fun and excitement, open to learning by listening to others, asking questions, and sharing ideas.

Curious people are not distracted by a fear of failure and have the courage to take a risk because the benefits of learning something new outweighs the drawbacks of feeling self-conscious or being seen as a novice in a new field. This makes them ideal collaborators because they are willing to step into collaboration and learn as they go. This is typical of an iterative or design-led approach to collaboration that is a golden thread that runs through this book.

Creating an environment where curiosity is nurtured is important for teams and organisations that innovate successfully [10]. They make experimentation part of the way things are done, and a failed experiment is seen, not as a costly mistake, but as an opportunity for learning. Stefan Thomke, William Barclay Harding Professor of Business Administration at Harvard University, notes that it is actually less risky to run a large number of experiments than a small number, because then even a low success rate will translate into a significant number of successes, which in turn diminish the financial and emotional costs of the experiments that are learning experiences.

Stefan Thomke also says that there is a human tendency to accept results that confirm our biases and view these as 'good'. However, results that go against our assumptions are often thought of as 'bad' and subjected to much more critique and scrutiny. Leaders have an important role in modelling a curious mindset by accepting that their views and assumptions can be challenged by data and experiments and taking this into account when making decisions.

Going to a meeting or a workshop about a topic that you know little about and that seems to be a long way from your discipline and viewpoint can feel risky, but if you can reframe it as being curious, you can see it as an opportunity to learn and grow.

Young children will ask lots of questions like 'why?' or 'why not?' as their natural curiosity pushes them to explore the world around them. These are deceptively simple questions that we can use to widen our own perspectives and confront our own assumptions. There is more about the power of great questions later in this chapter.

When we facilitate workshops, we begin by asking participants to be curious. It's a good reminder that it's ok to ask questions which others might find naïve and to challenge constructively. It's a prompt to stay open-minded, tolerant and especially to have fun. By doing this we want to reduce the barriers to taking part actively and to set the space up for exploration, creativity, and fresh thinking.

Being curious is not only important for collaboration; this mindset has been shown, by the recruiters Egon Zehnder, to be an important component of potential for performance improvement [3]. You can use these concepts to position yourself more broadly for your career. The other important factors of 'potential' are insight, engagement, and determination. High potential individuals go on to perform better in their roles than peers with lower levels of 'potential' thanks to their openness, growth mindset and curiosity.

2.1.4 How much do you know about your own mindset?

Here are some self-awareness prompts to help you start to explore your mindsets, approach and world views in relation to a potential or existing collaboration.

You may like to document these, in a journal, or use them as the basis for a conversation with a trusted colleague or friend. Remember the idea is to get a better understanding of your own perspectives and to go into a collaboration more aware of yourself. Be curious, honest and kind—there is no need to share these reflections with anyone else unless you want to.

- What are my values and purpose?
- What can I bring to this collaboration?
- What are my limits?
- Where do I like to collaborate?
- What are my self-interests?
- What are my unconscious biases [11]?
- What are my fears?
- How can I be most effective?
- How can I change?
- How will I choose to show up in moments of conflict?
- What can I do to narrow any divides?
- What do I need to do differently/next?
- How can I help myself feel comfortable sharing control?

2.2 'I should be so lucky'—the role of serendipity and how you can encourage it

Serendipity, a fortunate yet unplanned outcome, is an English word first coined in the late eighteenth century. 'The Three Princes of Serendip' were the stars of a fairytale in which the heroes 'were always making discoveries, by accidents and sagacity, of things they were not in quest of' [12]. Many scientific discoveries have been the result of serendipity, like stainless steel, post-it notes and even penicillin. It is often said that serendipity plays a huge underpinning role in innovation and why some companies are very keen to see their employees get back to a common workplace after the Covid-19 pandemic.

Serendipity isn't just luck though. It is luck combined with some wisdom, openness to opportunity, and tenacity. The good news is that it is possible to engineer more serendipity into your attempts to build collaborations using many of the skills that we will be talking about in this book. As French microbiologist and chemist Louis Pasteur once said,

'In the fields of observation chance favours only the prepared mind' [13].

No factor is more strongly associated with serendipity than a prepared mind, that is, the preparation to be open to opportunity and to spot it when it arises.

In our research for this book and in our own experience, we have heard many stories about the people who met in a lunch queue, at a conference registration desk or in a bar and then went on to have successful research collaborations. While serendipity was certainly at the root of these collaborations, the people concerned weren't simply lucky. They had put themselves in a situation where some luck, their willingness to engage with someone they didn't know, and a huge dollop of curiosity could come together to allow a serendipitous meeting of minds.

We've talked already about how relationships are the foundation upon which collaborations are built. If you think about your networks, personal and professional, you will see parts of the networks where there are really dense and strong connections and where a lot of people know each other. There will be other parts of your networks where the ties and connections are much looser—so-called 'weaker ties'. There will be people outside of your networks who you are not directly connected to but who are connected to people in your networks.

Serendipity tends to come from parts of your networks where you have weaker ties, because this opens you up to new perspectives, information and resources.

Ethan Zuckerman, an Associate Professor at MIT Media Lab, described engineering serendipity as, 'this idea that we can help people come across unexpected but helpful connections at a better than random rate. And in some ways it's based on trying to reassess this notion of serendipitous as lucky—to think of serendipitous as smart.'.

Figure 2.6. Engineering serendipity schematic.

You can help to engineer serendipity into your interactions with other people by taking some simple (but not necessarily easy) steps (figure 2.6).

2.2.1 Put yourself out there

Serendipity often arises when you meet people from parts of your network where your connections are relatively weaker and you have access to new perspectives, new information, and new resources. You can increase your chances of serendipitous meetings if you start to mix in different circles and learn different things.

It's worth looking out for opportunities that will give you the chance to put yourself out there, to meet people—particularly people that you might not have had the opportunity to meet before. People who sit outside your normal subject networks and boundaries. How could you do this? Conferences offer plenty of opportunities for networking. Webinars, forums, and workshops in subjects which you find interesting and are not directly related to your research have become much more accessible as a result of the move to online events during the Covid-19 pandemic.

2.2.2 Show up fully

If you've taken the time to attend an event, workshop, or conference, then to get the best return on this investment of time and allow serendipity a chance to show up, you also need to show up fully. This means being present and taking action to engage with the opportunities around you. It can be daunting to strike up a conversation with someone you don't know, but as the proverb goes, fortune favours the brave. Pull on your curious mindset, take an interest in people and have the courage to engage with someone new.

If you've been fully present at the event, this will be easier as you can use conversation openers about the previous session, or what you are both hoping to get out of being there. Once you've broken the ice and started to learn more about someone else, it's surprising how quickly you can find common ground … and then you are away.

2.2.3 Be a go-giver rather than a go-taker

It's always a more rewarding experience to be approached by another person that is genuinely interested in you, rather than someone that is interested in what you can do for them. Listen, and be generous with your connections and introductions. You never know—this person might not be your future collaborator but may have just the contact you need now, or at some point in the future. Exchanging emails, business cards or texts can be a great way of maintaining contact outside of an event and taking the relationship to the next level of trust. You can also create groups for people with common interests on a variety of social media platforms.

2.2.4 Send out a flare

Use social media and your networks to tell people about what you're working on, your challenges and your passions. You could ask your contacts to recommend people that you could speak to. Platforms like Givitas [14], cofounded by Adam Grant (author and Professor at the Wharton School of the University of Pennsylvania), aim to provide a way for people to ask for and offer help to each other.

2.2.5 Stay open and invest your attention wisely

Your attention is a limited resource. If you are to spot opportunities as and when they occur, you will need to be able to use your attention wisely.

In the fantastically named 1958 study, *The case of the floppy eared rabbits* [15] the authors identify that 'Because the methods and assumptions on which a systematic investigation is built selectively focus the researcher's attention, to a certain extent they sometimes constrict his imagination and bias his observations.'.

To increase your chances of serendipity striking you want to encourage your brain to pick up on nuggets of information that it might otherwise have filtered out as not relevant to the problem. Investing your attention in a broader range of possibilities, that is, keeping an open mind, is one way to do this.

An exercise Liz sometimes uses within a workshop is to ask a small group of randomly connected strangers, often deliberately seated next to each other, to come up with a research project they could conduct together. The aim is to push them to look for new connections and synergies and to allow serendipity to work its magic.

2.3 The key communication skills that make collaboration

If relationships are the foundation on which collaborations are built, then communication is the bedrock upon which the foundations are laid. Communication is the exchange of information between people, which occurs using language both verbal (spoken, written) and non-verbal (body language, visual, sign). We can support research collaboration by carefully using language, listening deeply and asking better questions.

2.3.1 Why does language matter for successful collaboration?

'Each of you possesses the most powerful, dangerous and subversive trait that natural selection has ever devised. It's a piece of neural audio technology for rewiring other people's minds. I'm talking about your language, of course, because it allows you to implant a thought from your mind directly into someone else's mind, and they can attempt to do the same to you …' [16]

Language is a key enabler for research collaborations. To appreciate the significance of language, it is useful to understand a little more about how language develops and how it is used.

We started learning languages from the day we were born. We learn the 'correct' way to speak 'our' language(s) and the words, sounds, movements and symbols associated with people, objects, and actions around us. The languages that we use are heavily influenced by the culture and context we are in. The philosopher, Ludwig Wittgenstein, said in his book, *Philosophical Investigations*, 'the meaning of a word is its use in the language'. Exactly how many words for snow there are in the Inuit language is hotly debated; however, what is clear is that there are many times more words for snow in Inuit than in English. The Inuit language is not alone in this—common words can mean very different things to different people in different contexts, and the same is true for the various disciplines in research.

This is demonstrated by a story that John Girkin, Professor of Optical Physics at the University of Durham, told us. Misunderstandings had arisen in a multi-disciplinary project he was collaborating on, due to different understandings of the word 'sensitivity':

'I was working with an analytical chemist, looking to develop a method for detecting chemicals in low concentrations in real-time as part of a production process. The project had regular bi-weekly meetings and after about 6 months it became clear that the two branches of physical science were travelling on parallel tracks.'

'I was very happy with the progress of our project at this point because the system could detect 1 molecule of the target chemical in 10^18 molecules. This seemed sensitive to me—but the chemist I was working with was concerned that the system was not sensitive enough.'

'Eventually one of us asked the simple question, 'what do you mean by sensitivity?' The physicist's perspective was that the sensitivity of the system is the lowest concentration of molecules that it can detect; whereas the chemist's view was that the sensitivity of the system is the combination of the lowest concentration that it can detect <u>and</u> measure what that concentration is.'

'Both are valid definitions of sensitivity, but the misunderstanding meant that the project was following two parallel tracks rather than an integrated approach. Once this difference in language was sorted out the system was modified slightly, which enabled a high quality, high impact project to ensue, with excellent publications and the installation of a system on an industrial micro-fluidic production line.'

We also asked 25 volunteers to describe what they thought of when they thought of the word, 'capacity'. The word cloud (figure 2.7) shows the diversity of meanings of this word among this relatively small group.

Julie Sanders, Professor of English Literature and Drama, and Deputy Vice-Chancellor at Newcastle University, told us that one of the best learning spaces of her career was a reading group at the University of Nottingham. Colleagues bought different things each week—an object, a text, a piece of music, and discussed them in a mixed disciplinary space. She said, 'I learned so much … and great projects came from it too.'

Figure 2.7. Word cloud derived from the responses of 25 participants to the question 'what do you think of when you think of the word Capacity?'

> **Quick write—an exercise in thinking about the multiple meanings of common words:**
>
> Take one of the words listed below, and write for 5 min (speed writing) everything you can think of about that word. Share.
>
> If you are facilitating and someone gets stuck move them on to another aspect of that word.
>
> Another way of doing this is to use an app, like Mentimeter, and ask your participants to input what they think of when they think of a particular word.
>
> | | | | | |
> |---|---|---|---|---|
> | Spin | Solution | Normal | Mean | Plane |
> | Noise | Slip | Observation | Volatile | Flux |
> | Attraction | Chaos | Integrate | Cleavage | Differentiate |
> | Scale | Sign | Rate | Gravity | Frame |
> | Relation | Relative | Moment | Term | Abstract |
> | Accommodation | Factor | Field | Force | Function |
> | Equal | Current | Energy | Present | Work |
> | Ground | Virtual | Common | Rare | Action |
> | Blue | Body | Capacity | Charge | Chance |
> | Duck | Order | Model | Refuse | Significant |
> | Natural | Transpire | Virus | Viral | Cope |
> | Culture | Play | Mole | Actor | Save |
> | Safe | Interest | Plant | Bond | Key |
> | Network | Goal | Bored | Marker | Power |
> | Web | Post | Trial | State | Stick |
> | Table | Value | Ruler | Sensitivity | |

2.3.2 What is the language of your tribe?

When you think about the languages you communicate with, you might think of your first language, the one you first learned to speak growing up. You may have learned other 'foreign' languages or be multi-lingual and you may have learned a dialect or words particular to places you have lived or even to your family. You have probably also learned to communicate in the language of your research speciality, your discipline or what we might call your 'tribe' and use the terminology, jargon, and acronyms widely used with colleagues and others sharing the same interests creating common and comfortable dialogue.

Just as a community of people defines its language through use and meaning, a community can be defined by its use of language. When everyone uses a common vocabulary, the community becomes more close-knit and comfortable with each other, and there is a greater sense of belonging and camaraderie. This had the evolutionary effect of fostering loyalty to our 'tribe'. People express their loyalty to a particular community or tribe through language, for example their loyalty to a nation, a football team, a religion, a subculture, or even a cult. Using the language of the tribe makes users feel like they are on the inside and being on the inside of a tribe is the safe place to be—it helps us to feel that we belong.

By analogy, as individuals study increasingly specialised topics, we learn the culture and language of our discipline. We become part of a social and academic 'tribe' associated with our discipline and specialisation and create another paradigm. We attend conferences with people that have similar interests and speak similar languages so we can immediately form bonds; we understand the private lexicon of the group including the terminology, jargon, acronyms, slang and even banter and 'in' jokes.

Academic in-jokes

How fast did the Moine thrust? About a mylonite. (c/o- Dave Petley)

sodium sodium sodium sodium sodium sodium sodium sodium sodium sodium sodium sodium Batman (c/o- Tony Ryan)

I used to be a post-structuralist, but now I'm not Saussure? (c/o- Chris Hewson)

Most expensive chemical in the body—ATP? (c/o- Mike Trenell)

A photon checks into a hotel and is asked if he needs any help with his luggage. He says, 'No, I'm travelling light!' (c/o- Ruth Patchett)

Using these tribal languages has had distinct advantages. It has enabled us to communicate fluently and efficiently with colleagues all around the world using the language of our discipline.

Like any other language, our specialist languages also adapt and change all the time, introducing new words and borrowing words from other languages and contexts to describe new findings and ideas.

Unfortunately, this efficient way of communicating can introduce some challenges when we want to communicate across tribal, or discipline, boundaries.

If we go on holiday or to live somewhere that uses a different language to us, learning some of the local language can help to build relationships with people that we meet. It can also help us start to understand and appreciate their culture. Exactly the same is true when trying to build research collaborations. The good news is that we can all learn new languages to a greater or lesser extent. Taking part in collaborative research is good training for speaking across disciplines and across institutions, it's a skill that you can learn; a muscle that you can develop with practice.

> **The 5 books of stupid: A technique for getting to grips with the language of other fields.**
>
> Former journalist and author Steven Kotler has a great technique called the '5 books of stupid', that he used to use as a journalist to get up to speed rapidly with new fields [17, 18].
>
> He starts with an easy to digest book, which might even be fiction, just to start to get to grips with some of the language. The second book is something like a popular science book, starting to bring in more of the vocabulary and concepts but nothing too heavy that would be frustrating. The next three books are more and more detailed, one might be a textbook, for example, and one might include trends or future thinking in the subject to help identify boundary conditions, state of the art, or topical challenges. At this point you might feel confident enough to tackle some research papers in a different field.
>
> What we like about this progression is that it lowers the bar to entry, to starting to learn the language of a new field bit by bit. And it doesn't have to be books; there are lots of resources available online including podcasts, TED talks, documentaries and films that could help you to ease your way into the language of a new field in a less daunting way.

One of the first steps in multidisciplinary collaboration is often to create a new common and inclusive language or lexicon to facilitate communication and reduce misunderstanding.

Peter Galison, Professor of the History of Science and of Physics at Harvard University describes a concept, borrowed from anthropology, of a 'trading zone' [19] where two different cultures come together. Within the highly localised context of the trading zone the two different cultures can collaborate.

Peter Galison also says that to move the whole discipline of physics forwards, theoretical and experimental physicists must collaborate with each other. To do so they have got to establish common ground and a shared understanding of what they will do together. They also need to establish what the words they will use with each other mean.

In their collaboration, or trading zone, the terms that they use become simplified and a proto or 'pidgin' language emerges. The purpose of the pidgin language is to facilitate collaboration. Terms may be used by the collaborators in ways that are different to how they would be used in their 'home' discipline and so the trading zone, or the collaboration, becomes a special place with its own rules.

We have often witnessed this first-hand in workshop discussions when groups can spend time discussing the semantics and meaning of words. One example of this is 'sustainability', which can have many different connotations depending on the context and discipline. This can sometimes make for difficult discussions as each interpretation 'fits' its corner. Where we identify these issues in advance, we clarify the definitions we are going to use, with the participants, at the start of a workshop. This helps to create a common focus and enable the conversations to begin.

In a workshop Liz facilitated for the research charity Autistica and the Alan Turing Institute, it was important to establish a common understanding of what was meant by 'Artificial Intelligence (AI)' as well as a common understanding of what was meant by 'early interventions in Autism'. Building in time during the workshop for more 'tribal' conversations enabled easy discussion of complex, specialist issues. The conclusions from these were reported on to the whole group, then built on further in multi-partner conversations. By enabling exchanges to happen, understanding and trust began to develop between the AI researchers, Autism researchers, and people with lived experience. The outputs were co-produced and developed further. The outcome was AI driven solutions which have gone on to have a positive impact on the lives of people who live with Autism.

All this building of language takes time, effort, and energy. Sometimes, especially when we are busy or tired, this feels like just too much and it feels so much easier to stick with what and who we already know. 'Left to their own devices, people will choose to collaborate with others they know well—which can be deadly for innovation.' [20]

When we have brought new groups of people together, we have observed that when individuals arrive there is tendency for them to self-organise around tables of their 'own'. This is a perfectly normal human behaviour, so if we want to establish new collaborations it is the facilitator's role to get the participants mixing and mingling.

At the start of your collaboration, you might find it useful to go through this process of clarifying of, 'what do we might mean by …?'. You could start to reflect on the jargon that your discipline uses and whether any of those terms have 'double meanings'.

Facilitators often deliberately set up conversations between people who come from different backgrounds. As conversations begin and develop, people gain an understanding of each other's perspectives, problems are sometimes solved, and we begin the process of building this common understanding—the start of a common language.

It's music to our ears when we hear that workshop participants who had never met before identified a way to solve a problem and are going to collaborate. A new tribe is emerging.

2.3.3 Language as barrier—how we use language to exclude as well as include

Language, that amazing superpower, that efficient communication tool, also has some down sides if not used carefully. We can use it in ways that create 'closed' communities which excludes others, deliberately or unintentionally.

'It seems that we use our language, not just to cooperate, but to draw rings around our cooperative groups and to establish identities, and perhaps to protect our knowledge and wisdom and skills from eavesdropping from outside. And we know this because when we study different language groups and associate them with their cultures, we see that different languages slow the flow of ideas between groups. They slow the flow of technologies.' [17]

Historically, the suppression of language has been used to oppress minorities, their cultures, and traditions. This is just one way that language can be used to exclude people that are different.

Hopefully, we don't see such extreme use of languages to oppress others within our universities, but we have both observed people use subject specific jargon and language in workshops that 'locks' other people out and reinforce the different tribes that they belong to. We need to make sure that there is enough time in workshops and in our interactions with others to explain jargon and terms which may not be familiar to all participants.

This is particularly important when we are co-creating and co-producing with community representatives or the public who might already find universities intimidating and exclusive. If, on top of that, language is used which they feel 'locks them out' of conversations it will be even more difficult to build the trust-based relationships which will be needed for full participation in a research project.

2.3.4 Tips for broadening and clarifying your language

2.3.4.1 Build your network and have lots of conversations
If you want to build strong relationships with others from different backgrounds, and avoid misunderstandings, it's good to practice talking to lots of people with different backgrounds. The more we use language, the more we become used to articulating concepts and meaning. We learn to be more accurate and clearer in our communication. We get to know which terms in our own lexicon, or tribal language, also have other meanings, and our awareness of multiple meanings of other terms also increases.

2.3.4.2 Use images as well as words to clarify meanings

'When we communicate with others, psychological research shows, we are often too indirect and abstract. Our words carry more weight if we are more concrete and provide vivid images of goals.' [21]

The inclusion of pictures, or visual cues can help develop shared understanding more quickly. Take, for example, the word 'goal'. What 'vivid image' does that conjure up for you?

Perhaps it is that of a football (soccer) goal, or a set of rugby posts. Other goals and sports are also available!! To score a goal in football or rugby it will help to know which side of the crossbar you have to place the ball.

Any misunderstanding about the type of goal could be quickly resolved with a picture of the goal you are trying to score into (figure 2.8).

We're quite used to including pictures and diagrams to help enrich our written communications, but they can really help with verbal conversations too.

Figure 2.8. Which 'goal' do you mean? Images by Pascal Swier (@pascalswier16) and Aline de Nadai (@alinedenadai) for Unsplash and Victor Dernitz for Pixabay.

2.3.4.3 Find role models
Watch 'good' communicators and collaborators at work. What makes them a good communicator? What values underpin the way they behave?

2.3.4.4 Learn from your mistakes
Reflect on those conversations you have had when things haven't quite gone the way you expected them to go ... why did this happen?

You might find it useful to reflect on a situation where you said, verbally or via email, something which you thought was straightforward, but which resulted in a difficult conversation simply because there had been a misunderstanding or misinterpretation. How might you have approached this differently?

2.4 Listening and its role in collaboration

'Listening, unlike other senses, is necessarily collaborative, an act of sharing and so a powerful means to create and reach materially different and more ambitious outcomes than those allowed simply by staying within our own thoughts.' John Keith—Managing Director—Head of Financial Institutional Coverage at BNP Paribas

When was the last time you felt genuinely listened to by a colleague? Hopefully it was recently, as we all need to feel respected and listened to. Unfortunately, our time and attention are under constant pressure with noise that pervades every part of our lives and its surprisingly rare to feel genuinely heard.

'Today people resist listening to one another. They find face-to-face encounters trying and phone calls intrusive, preferring text and wordless emojis. Many people refuse to listen to opposing views, shouting down or walking out on speakers who challenge their thinking. If people are listening to anything, it's probably through earbuds, safe inside their own curated sound bubbles.' [22]

We have explained that collaborations are built on relationships; but what kind of a relationship can you have with someone that does all the talking and never listens? One of the most sure-fire ways of finding people we want to work with is to listen generously and with curiosity.

Although it is essential to have people with different perspectives in the same room, that alone it is not sufficient to enable collaborations to form and develop. We need to listen deeply to and learn from each other. Otherwise, we may as well just be back in our silos. The more diverse range of people you listen to, the more insights you will learn, the better your understanding will be and the more ideas for solutions you will be able to generate.

Listening is the basis of connection and hence relationship and collaboration, yet it is a skill that is not widely taught. Indeed, many academic researchers have been trained to communicate solely in terms of putting forward and defending their own hypotheses and opinions. Listening skills are like a muscle—if you don't use it, you lose it. Some people have more of a natural talent for listening, but everyone can improve their listening skills with practice. To develop your collaboration skills, you can invest time in improving your listening skills. Even if this doesn't come naturally, it could become your superpower.

As long ago as 1957, researchers at the University of Minnesota were studying the challenges of listening in business [23], and found that it's actually harder to concentrate while listening than during any other form of personal communication. This is because humans think much faster than we talk. The average rate of speech is around 125 words per minute, but our brains process this information almost four times as quickly [24].

This creates a listening 'gap'—which mean that the listener literally has spare capacity for thinking when they listen. Taking notes, especially handwritten notes, helps us to concentrate on what is being said by making use of some of that spare processing capability. The process of listening, summarising, and writing prevents our brains from racing ahead and fosters understanding of what is being said [25]. Interestingly, students that took notes on laptops during this study tended to transcribe what was being said and retained less.

Of course, it's not always possible to take notes, so it's important to have some other tactics for ensuring that we use that spare processing capability constructively otherwise our attention may drift. To listen deeply we need to be fully present in the moment.

2.4.1 Models of listening

There are many different models for listening, often described as 'levels of listening' that can be found in the literature. One that we have found to be particularly helpful is that described by Julie Starr, a widely respected authority on coaching, mentoring, and personal development [26]. In this model, as your focus and attention on the speaker increases, your listening level 'deepens' (figure 2.9).

1. **Cosmetic listening**—pretending to listen. Looking like you are listening, nodding, 'Hmmmming', but your attention is elsewhere.
2. **Conversational listening**—you are actively involved in the conversation, listening, thinking, talking etc. Your mind sometimes drifts away, either prompted by other things that are happening or something that the speaker has said. You might just tune out because your brain is using its spare processing power in ways which may, or may not, be helpful.
3. **Active listening**—you are very focused on what the other person is saying. You may be using some of your brain's processing power to take notes, ask clarifying questions, paraphrase and reflect back.

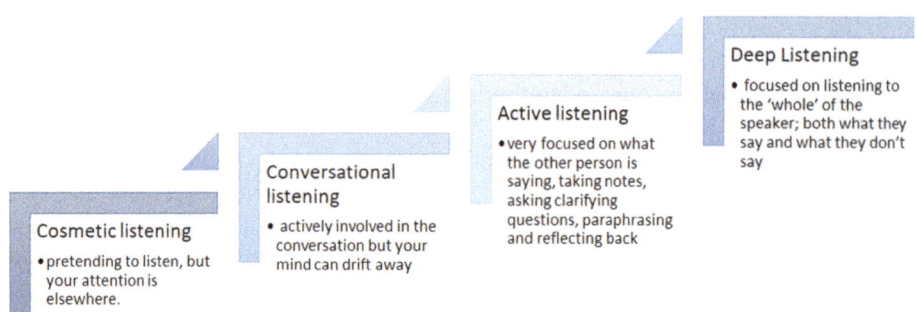

Figure 2.9. Summary of the levels of listening described by Julie Starr [26].

4. **Deep listening**—you are focused on listening to the 'whole' of the speaker; both verbal and non-verbal communications (what they say and what they don't say). You can pick up on the feelings and emotions of the speaker. This type of listening enables the speaker to feel safe and understood—and is more likely to lead to a meaningful relationship between the speaker and listener.

2.4.2 Non-verbal listening

It might not seem immediately obvious that non-verbal communication is important in research collaboration.

In actual fact, humans have evolved over hundreds of thousands of years to communicate very efficiently through non-verbal signals. These signals carry huge amounts of information, particularly about whether someone feels uncomfortable, which can give a clue as to whether or not they are likely to want to form a relationship or collaboration.

This is one reason why face-to-face communication is so important; devices can get in the way of these subtle signals. Non-verbal cues give us insights into the feelings and emotions of others. They can help us to empathise more effectively.

For example, you can tell a great deal about someone's emotional state by listening to where in the body the words are coming from. Is the person taking big belly breaths and speaking confidently, or are they speaking from the throat with tension or nerves? What is their posture like?

In an entertaining and fascinating talk available online [27], Joe Navarro, a former FBI agent and expert in non-verbal behaviours, describes some key indicators that we all display if we are feeling unsafe or something is bothering us. These include, but are not limited to nail and lip biting, wringing hands, pursed, or sucked in lips, nose wrinkling and eye-blocking. Some of these behaviours are millions of years old and are deeply rooted in our subconscious. In a face-to-face workshop, either facilitators or participants might pick up if someone else was giving off these signals and could then take steps to find out what was troubling them.

One cause of so-called 'Zoom fatigue' is that during video calls we have to work harder to process non-verbal cues like facial expressions, the tone and pitch of voices, and body language. It takes a lot more effort for our brains to communicate when they can't make use of the non-verbal signals which we are so well adapted to interpret. The increasing use of digital technologies in collaboration is something we're going to come to in the very last chapter of the book—and using them effectively for listening and communication will be key for future success.

2.4.3 Supporting listening in diverse groups

One point to note here is that some neurodiverse researchers may experience challenges with picking up on non-verbal cues. Neurodiversity is a term used to describe a range of neurological characteristics, including autism and dyslexia. People with hearing difficulties may have challenges listening to verbal conversations, especially in a room which becomes noisy.

We will pick up on the importance of diversity and inclusion in the next chapter. Given the subject of this section is listening, we wanted to highlight here the importance of checking in with all of your participants, prior to any workshop or meeting, about any additional needs they might have. You can then take the steps needed to make sure that everyone can contribute fully to the discussions.

2.4.4 Asking the right questions—the key to collaboration

'I have learned that everyone is interesting if you ask the right questions.' [22]

A core part of listening and a curious mindset is asking questions. We have seen time and time again that often apparently simple questions can get right to the heart of an issue and can challenge long-held assumptions or 'sacred cows' which might have become outdated.

Author Oscar Trimboli explores the potential of what he calls the '125/400' rule; the gap between the speed at which we speak and which we think, in his book, *Deep Listening* [28]. He describes how the brains of speakers, as well as listeners, are thinking faster than they can speak. Listeners can help speakers to express their thinking more fully by asking the right kind of questions. When exploring unmet research needs, if we can help our collaborators more fully express their challenges and experiences, we may uncover new directions for research which were not immediately obvious, or constraints which hadn't previously been communicated.

The intention behind asking the question is important. As Pilar Orti of the '21st Century Work Life' podcast says so beautifully:

'Ask thoughtful, insightful questions—especially questions that you don't know the answer to. Ask questions when you want to find out what someone else thinks, not when you want to prove a point, and definitely not when you're being sneaky about people buying into your decisions.' [29]

When you get deep into a subject or area of expertise, you can stop questioning assumptions and start taking things for granted. You don't even know that this is happening to you—like a fish that doesn't know it's in water. This concept is perfectly illustrated by this quote from the late author and professor of creative writing, David Foster Wallace's 2005 speech 'This is Water' [30]:

'There are these two young fish swimming along and they happen to meet an older fish swimming the other way, who nods at them and says 'Morning, boys. How's the water?' And the two young fish swim on for a bit, and then eventually one of them looks over at the other and goes 'What the hell is water?"

So, what are the 'right' sort of questions for getting into collaboration? Firstly, questions need to be genuine, as we've already said. Open questions are much better for encouraging a dialogue than closed questions—which elicit a yes or no answer. Questions are good way of opening up a conversation and getting to know a potential collaborator as a person as well as a researcher. Ask about their interests, knowledge, and motivations.

If you want to get into a research collaboration, you're going to need to go beyond the 'getting to know you' questions. When you do that, you want to move from asking questions about the person to asking questions about the idea, problem, model, story, or perspective. As well as helping you to have more understanding, asking great questions can help others challenge their own assumptions, reach new insights, and get them thinking about their own thinking. These sorts of questions are called reflecting questions. When you ask a reflecting question, don't rush the other person into an answer—treat silence as thinking time. Likewise, when they have finished speaking—don't jump right in, in case they haven't finished or have something else to add.

One way to help people think about their challenges in a positive and solution driven way is to use the framing question, 'wouldn't it be great if we could …'

If you want to go beyond a 'top of head' answer, a great tip from business coach Michael Bungay Stanier [31] is to ask, 'and what else?' or, 'tell me more …'. Oscar Trimboli explains in his book, *Deep Listening*, that when you use phrases like 'tell me more' it literally slows down the conversation, so you give someone the opportunity to align their thoughts more clearly. They have time to think through the idea, and identify the most important themes to bring out, and their response will be more nuanced and richer as a result. It really helps to extract even more information, which might turn out to be vital for identifying an innovative solution.

You might be worried that by asking questions, potential collaborators or other people in a workshop might judge you for asking questions. As long as your question is genuine, you have no need to worry. In *Rebel Talent* [6], Francesca Gino described research that shows that people think of us as smarter when we ask questions, than when we don't. She explains how when we ask genuine questions and show an interest in other people and their ideas, trust deepens, and our relationships become more interesting and more intimate. What's more, when you ask questions of others, they share more, and they reciprocate. They ask questions of you, which sets up a positive spiral of give and take that strengthens the connection between you—the start of a beautiful collaboration.

> **Reflecting questions**
>
> - Wouldn't it be great if we could ...?
> - What would make the biggest difference?
> - What would be your ideal scenario?
> - What is the real goal here?
> - How might we ...?
> - How long have you been thinking about this?
> - What is the model that you are using?
> - What is important to you at the moment?
> - What if it was easy?
> - What help do you need?
> - What do you think needs to happen here?
> - What are our strengths?
> - Where do we sit in the pathway to impact?
> - What does [this new perspective] mean/tell us?
> - And what else?
> - Tell me more ...

2.4.5 Using a soapbox to create a space for better listening

In workshops and events, it's relatively common to observe the behaviour of 'grandstanders'—those who are very keen to speak and less keen to listen. Behaviours like these inhibit collaboration as they send out signals that the 'grandstander' is not interested in what is important to everyone else. Not listening can destroy the rapport of a group very quickly.

As a facilitator, it is important to think about how best to intervene when you observe these kinds of behaviours. Some of the more confident and secure workshop participants may make their own form of interventions. Sometimes these 'grandstanders' just need an opportunity to say their piece and then they can relax.

Liz has found that one successful way of giving everyone an opportunity to make their point uninterrupted is through a short, managed timed input or 'soapbox'.

> **How to run a Soapbox**
>
> Soapboxes are 2 min strictly timed 'inputs' by an individual. They give everyone attending a meeting or a workshop a chance to have their say in a managed way and contribute constructively to the debate and content of the meeting. Don't encourage the use of slides and don't advertise the soapbox opportunity before the meeting, to ensure that the input is spontaneous.
>
> When running workshops online, you may wish to restrict a soapbox to 1 min in duration, and to use the 'chat' function as the way of people 'signing up' for a soapbox.
>
> Soapboxes often work well where you know that there are likely to be a number of people that will feel happier having had the opportunity to express their views and ideas, or if someone wants to tell the group about an upcoming event or other resource relevant to the subject of the workshop.

2.5 Key learning points—chapter 2

- Preparing yourself to collaborate is a good way to position yourself for serendipitous encounters.

- Proactive, abundance and growth mindsets support collaboration; defensive, scarcity, and fixed mindsets make it more difficult.

- Curiosity is part of a growth mindset and is a key behaviour for collaboration.

- A failed experiment is an opportunity for learning.

- Serendipity is luck combined with wisdom, openness to opportunity and tenacity.

- The parts of your networks with weaker ties are more likely to lead to new perspectives, information, and resources.

- To encourage serendipity, put yourself out there, be a go-giver, send out a flare, stay open, and invest your attention wisely.

- Research collaboration is facilitated by carefully using language, listening deeply, and asking better questions.

- Language can be used to exclude as well as include.

- To broaden and clarify your language: build your network, use visuals, find role models, and learn from your mistakes.

- Listening is vital to learn from each other.

- Verbal and non-verbal communication are both important for establishing relationships and collaborations.

- The right questions are open, genuine, and reflecting.

- Asking genuine questions builds trust.

- Sometimes you have to create a space for people to 'have their say' to create a space for listening.

References

[1] https://wellcome.ac.uk/sites/default/files/research-careers-tips-running-research-group-2018-05-17.pdf

[2] Plenty R and Morrissey T 2020 *Uncertainty Rules? Making Uncertainty Work for You* (Cork: Atrium/Cork University Press)

[3] Gino F and Staats B 2015 Why organisations don't learn *Harv. Bus. Rev.* 112–18 (November)

[4] Brown B 2012 *Daring Greatly: How the Courage to be Vulnerable Transforms the Way we Live, Love, Parent and Lead* (New York: Gotham Books)

[5] https://firstround.com/review/the-remarkable-advantage-of-abundant-thinking/

[6] Gino F 2018 *Rebel Talent—Why It Pays to Break the Rules in Work and in Life* (New York: Harper Collins)

[7] Dweck C S 2013 *Mindset* 7th edn (London: Robinson)

[8] Moser J S, Schroder H S, Heeter C, Moran T P and Lee Y-H 2011 Mind your errors: evidence for a neural mechanism linking growth mind-set to adaptive post-error adjustments *Psychol. Sci.* **22** 1484–89

[9] https://timeshighereducation.com/blog/your-biggest-asset-academic-career-success-growth-mindset

[10] Thomke S 2020 Building a culture of experimentation *Harv. Bus. Rev.* (March–April)

[11] The resources here may be helpful https://implicit.harvard.edu/

[12] https://lexico.com/definition/serendipity

[13] Liestman D 1992 Chance in the midst of design: approaches to librarian research serendipity *Res. Q.* **31** 524–32

[14] https://giveandtakeinc.com/givitas/

[15] Barber B and Fox R C 1958 The case of the floppy-eared rabbits: an instance of serendipity gained and serendipity lost *Am. J. Sociol.* **64** 128–36

[16] Pagel M 2011 TEDGlobal

[17] https://forbes.com/sites/stevenkotler/2012/09/04/how-to-learn-anything-in-five-not-so-easy-steps/?sh=2777007628c8

[18] https://successpodcast.com/show-notes

[19] Galison P 1999 Trading zone: coordinating action and belief (1998 abridgment) *The Science Studies Reader* ed Mario Biagioli (London: Routledge) pp 137–60

[20] Ibarra H and Hansen M T 2011 Are you a collaborative leader *Harv. Bus. Rev.*

[21] Gino F 2019 Cracking the code of sustained collaboration *Harv. Bus. Rev.* 73–81

[22] Murphy K https://theguardian.com/lifeandstyle/2020/jan/25/its-time-to-tune-in-why-listening-is-the-real-key-to-communication

[23] Nichols R G and Stevens L A 1957 https://hbr.org/1957/09/listening-to-people

[24] Carver R P, Johnson R L and Friedman H L 1971 Factor analysis of the ability to comprehend time-compressed speech *J. Reading Behavior* **4** 40–9

[25] http://collective-evolution.com/2016/06/05/a-secret-for-learning-dont-take-notes-with-your-laptop/

[26] Starr J 2008 *The Coaching Manual—The Definitive Guide to the Process, Principles and Skills of Personal Coaching* (London: Pearson)

[27] https://cmxhub.com/video-joe-navarro-the-power-of-nonverbal-communications/

[28] Trimboli O 2019 *Deep Listening*

[29] https://virtualnotdistant.com/podcasts/psychological-safety?rq=psychological%20safety

[30] Foster-Wallace D 2009 *This Is Water: Some Thoughts, Delivered on a Significant Occasion, about Living a Compassionate Life* (New York: Little, Brown and Company)
[31] Stanier M B 2015 *The Coaching Habit: Say Less, Ask More & Change the Way Your Lead Forever: Say Less, Ask More & Change the Way You Lead Forever* (Vancouver: Page Two)

IOP Publishing

Research Collaboration
A step-by-step guide to success
Annette Bramley and Liz Ogilvie

Chapter 3

Leading by example: preparing your team to collaborate

Collaboration and innovation have both been described as 'contact team sports'. It's a good analogy—rugby is a team sport where players with different specialisms come together—and literally collide with each other! Collaborative research collisions happen when people with diverse talents, interests and expertise meet, perhaps at a workshop, or a conference or even on twitter. After some (what might seem like random) conversations, ideas for research collaboration emerge and a team is formed to take these ideas forward. In fact, Florida State University in the USA has a series of interdisciplinary networking events called 'Collaborative Collisions' [1] with the goal of bringing people together from across their campus.

A key feature of collaboration is bringing a multiplicity of ideas and perspectives to address a problem. As research becomes more specialised, larger teams are required to make progress, even within single disciplines. In this chapter we'll highlight the benefits of making sure there are diverse perspectives, particularly for challenge-led research collaboration.

In chapter 1 we looked at the difference between collaboration and teamwork, and there we alluded to the importance of leadership in successful research collaboration. Here we will explore this in more detail and look at the leadership behaviours that support research collaboration, illustrating this with case studies. We'll also look at the role leaders have in creating the kind of environment where collaboration can thrive, the importance of a concept called 'psychological safety' and the impact of power dynamics on collaboration.

In a blog for the *Huffington Post*, former advertising executive turned entrepreneur Lynn Power offers some prompts for thinking about collaboration [2].

- Do we have a true diversity of thought, and have we created an environment where everyone can participate as an equal? (Diversity of background, role, gender, experience, culture, social status.)

- Is there a hidden bias towards the person who is leading the project as the most 'senior person'? And is this inhibiting true collaboration?
- Are we able to stretch our team's thinking and encouraging equal participation from all points of view?
- Are we acknowledging and respecting everyone's time and contribution giving everyone credit and ownership?
- Have we been clear about the expectations of everyone's role and outcome, so that people aren't participating with false expectations in a process that's unsatisfying for all?

We hope this chapter will help you to see why these issues are important and give you some tools for creating the kind of team environment that will enable your collaborations to be successful.

3.1 Diversity and collaboration

'When people from a singular background are placed into a decision-making group, they are liable to become collectively blind'—Matthew Syed [3]

It has been shown again and again that diverse leadership teams and diverse workforces are proven to bring the best returns on investment. Entrepreneur and journalist Shane Snow, in his book *Dream Teams: Working Together Without Falling Apart* [4] describes multiple dimensions of diversity:

- **Cognitive diversity** = different ways of thinking
- **Demographic diversity** = different demographics of people
- **Racial diversity** = people of different races
- **Gender diversity** = people of different genders
- **Age diversity** = people who grew up in different generations
- **Personality diversity** = people with different ways of rolling

As researchers, we need more than ever before to be able to embrace diversity and difference—whether this be disciplinary, gender, career stage, social status, or culture. We need to be able to work with conflict and creative tension; to be curious ourselves and encourage others to be curious.

Mary Stuart, who retired recently as the Vice-Chancellor of the University of Lincoln, highlighted to her colleagues the importance of avoiding group-think for good research. She told them:

'If we want good science we can't have group-think'; and

'We want to do interdisciplinary research [at Lincoln] because it avoids group-think, and because it's exciting.'

As the most senior leader at her university, Mary was setting out her vision for high quality research; research that avoids group-think and pushes at boundaries, and to do that in a university-wide setting means drawing in perspectives from all over the research base, nurturing interdisciplinary research.

Diversity in collaborations takes many forms—gender, nationality, culture, disciplines, place, social status, and lived experience to name just a few. Diverse collaborations take time to flourish and for trust to form. It takes time to understand the language and nuances that each contributor brings. We need to be able to have those conversations that enable us to learn more about each other's 'worlds'; how different 'tribes' think and understand an issue; how different disciplines have different understandings of what constitutes 'an answer' or 'a solution'.

Former Olympian and author Matthew Syed has explored how important 'cognitive diversity' is to problem solving in his book, *Rebel Ideas*. This view is supported by many other books and studies. Shane Snow explains that visible characteristics like gender and race are often pretty good indicators of probable cognitive diversity. Probable is an important qualifier, because researchers Alison Reynolds and David Lewis have found that a team that is diverse in age, gender, and ethnicity, but who all had the same thought process lacked enough unique perspectives to solve complex problems [5]. They say:

'These were PhD scientists who had been attracted to biotech to explore their specialties. But, with little cognitive diversity, they had no versatility in how to approach the task.'

Diversity can also introduce friction between participants that needs to be harnessed for its creative power rather than its destructive power. This can be difficult in teams and organisations where conflict is traditionally avoided rather than channelled. We need to work together to create a climate and a culture where this can happen, whether that is within a workshop or within a university. Trusting each other and listening to each other without judgement to understand different perspectives and ideas is essential. This is how relationships get built and collaborations are formed. This type of culture is called a psychologically safe environment, and we will come back to this later.

3.1.1 Team research and lone scholarship

One of the critiques of the approach to 'team research' that has been levelled at us over the years is that in some parts of the research base it is possible for a lone scholar to undertake a programme of research which straddles two traditional disciplines. In some way, the individual themselves is then inherently interdisciplinary. This is certainly true in some areas of research where individual scholars undertake a really important and valid type of research. Diversity of approach and method within our research base is important for its health and its vibrancy, just as it is important for teams to be diverse. We need to have the curiosity, understanding, and intellectual tolerance for all kinds of approach.

We would, however, reflect that it is also hard for a lone scholar to avoid groupthink with themselves, to bring a diversity of life experiences and perspectives to the research alongside their professional knowledge and to have the capacity to take their ideas forward as rapidly as a team working together. There is a place for lone scholars and for team research within our research base and for people to move in and out of collaborations at different times in their careers.

If you think about it, the humanities scholar producing a monograph (which implies that they work alone), has actually worked with a team which might include librarians, archivists, galleries and museums, performing arts companies or other scholars to produce that work. The acknowledgements pages in these sorts of books often tell a story of diverse collaboration. We hope that those who see themselves as lone scholars, wherever they sit in the spectrum of research disciplines, will also find ideas of interest in this book to support them in engaging in collaborations.

3.2 The role of leaders in research collaboration

Why do we often see that the success of a research collaboration is so dependent on those that lead the team and their leadership skills? What can a leader do to increase the chances of a collaboration being successful? These are two questions we've set out to answer in this part of the book. You can also use these insights to help you work towards being the leader you want to be, or wish you had.

Leadership is a key ingredient in making collaboration happen. It's the catalyst that brings people together behind a shared cause. It's the fuel that sustains the collaboration even when there are challenges. Morten Hansen, management professor at the University of California Berkley said in a *Harvard Business Review* article co-written with Hermina Ibarra, Charles Handy Professor of Organizational Behaviour at the London Business School that, '[Leaders] not only need to orchestrate the conditions for employees to collaborate on the right things, they need to show a strong hand in guiding collaboration efforts' [6].

Picking up on the analogy of 'orchestration', let's think about a research collaboration as an orchestra, and the team leader as the conductor of that collaboration. They do not teach their 'orchestra', i.e., their research team, to do research, but provide them with the environment to perform at their best. They guide and encourage them. Superstar orchestra conductor Gustavo Duhamel said, 'I think it's a very important collaboration between the conductor and the orchestra—especially when the conductor is one more member of the orchestra in the way that you are leading, but also respecting, feeling and building the same way for all the players to understand the music'.

Similarly, the value of great multidisciplinary research leadership to research is a bit like the value of a conductor to an orchestra. Troy Peters, conductor of the San Antonio Youth Orchestra, explored how to lead and collaborate like an orchestra conductor in his entertaining 2016 TEDx talk [7]. The Dutch conductor Jules van Hessen also considered this theme in his 2017 TEDx [8]. While orchestras can play without a conductor, Troy Peters demonstrates how an orchestra's conductor can have a measurable impact on the quality of the sound it produces. The vision for the piece comes from the composer and is set out in the musical score, but the composer is not leading the orchestra. The musicians may be brilliant at playing their individual instruments—the best even in the world—but their role is to play and not to conduct.

The conductor has the responsibility of interpreting the notes on the page and sharing the vision with the orchestra. They make the most of the skills of the musicians, blending and balancing the sounds to deliver the music with the pace and mood that the composer envisioned. They pay attention to the details while not losing sight of the overall vision.

They inspire the passion and commitment of the orchestra and maintain the relationships within the team. They help members of the orchestra listen to each other, and help the audience listen to the orchestra. They do not teach the musicians how to play but trust them to play to the best of their ability. They take care of the environment so that the musicians are ready and able to execute. The role of the conductor is to enhance the sound of the whole, and not to get in the way.

In a multidisciplinary research team, researchers come together, often to investigate a specific problem which they believe to be important. The role of the leader of the project, or of a workstream, is to create an enabling environment and a culture which will enable people to give of their best and feel confident in sharing ideas. The role of the collaborative leader is not to lead all the workstreams themselves but to make sure everyone understands the vision and that the right degree of blending and sharing happens between researchers and workstreams.

The best leader, or leaders, are not necessarily the world-leading researcher(s) in any single one of the research streams but are skilled in managing relationships, solving problems, and motivating and inspiring others.

Leaders of collaborations need to optimise the balance between individual interests and delivering the project as a whole. To do this they may have to make hard decisions, for example, about how resources are used or if a line of inquiry needs to be stopped or redirected. They will be challenged and need to allow debate yet know when to bring unproductive discussions to a close.

In his book, *The Infinite Game* [9], author Simon Sinek talks about the importance of embracing what he calls an 'infinite mindset'. Amongst other things, adopting this type of mindset means imagining and making the future happen by working effectively with others, inviting expertise and contributions from specialists at relevant times in the life of a long-term research programme.

In an infinite game, and in some research collaborations, team members tend to contribute different skills at different points in the programme, and to bring in additional skills as the programme develops. This fluidity and flexibility of approach also means that teams are more able to respond to unexpected directions or new avenues as well as bringing new perspectives to difficult problems.

The leader of the collaboration, or orchestra, has the key role of integrating what we might think of as a guest soloist, into the team. They also need to maintain the engagement of core team members with the overall project even when they may not have such a large part to play, so that they do not become disenfranchised and continue to feel part of the team. The leader has a position of formal and informal power, and the best collaborative leaders are comfortable sharing that power and using it for the interests of the collaboration, to create a collaborative culture.

3.3 Behaviours of people with good leadership skills

It can be unhelpful to label people as 'good leaders' or 'bad leaders'. Organisational norms and culture can play a large part in the relative perception of someone's leadership. You can be a good leader in some situations and not so great at other times. Both introverts and extroverts can make fantastic collaborative leaders, although our working culture may require and/or reward extroverted behaviours, especially for those in leadership positions.

It is more helpful to think about the kinds of skills that someone with good leadership in a specific context would show. So in the specific context of multi-disciplinary research collaboration, some behaviours, skills and attributes of someone that is likely to be able to lead successfully include:

- Honesty
- Humility
- Unbiased
- Unpretentious
- Integrity
- Inclusive
- Curious
- Generous
- Patient
- Steadfast
- Power-sharing
- Tactful
- Appreciative
- Intellectually tolerant
- Fierce professional will
- Able to exercise self-restraint
- Empathetic perspective
- Listening to understand
- Respect for the contributions of others
- Quiet in your own voice
- Able to bring people together
- Commitment to the cause
- Absence of arrogance
- Commitment to partnership working
- Clarity of thought and message
- Brokerage and negotiation
- Building on and experimenting with the ideas of others
- Being willing to have your ideas and opinions challenged constructively
- Encouraging feedback and working with it
- Taking time to understand others and to encourage others to understand each other
- Making others feel at ease so that they can bring forward ideas
- Delegating and empowering others with responsibility
- Confidence to be who you are and share this with others
- Able to make practical judgements and weigh up views/trade-offs

It's quite a list isn't it?! If you aspire to lead, the good news is that while some of these skills may come to you naturally, others can be improved with practice, like a muscle. If you are trying to put together a research collaboration, it's a good mental checklist to run through when thinking about who the individual(s) are that are best placed to take leadership positions within your team.

Being a research leader in a collaborative team, particularly a multidisciplinary one, is far more demanding on the time of an academic than conventional research. This is another reason why it might not be in the best interests of a project to have the world-leading researcher in the position of a research leader. World-leading researchers already have many pressures on their time, and a high degree of power and status arising from their research track record. Others will defer to them which may not lead to the best decisions being made, or the most creative avenues being pursued.

Abby Wambach, the former captain of the US women's soccer team and two times Olympic Gold medallist says in her book, *Wolfpack* [10], that, 'Leader is not a title the world gives to you, it is an offering you make to the world'.

A research leader serves the team that they lead. They need to invest time to get to know others in the project team and to consult with colleagues. The nature and implications of the choices they have to make may mean that established protocols are not always useful, and decisions not easily delegated.

They will have an advocacy role for others and will need to be prepared to develop the talent and skills of those within their team. They'll need to be a role model, setting the tone and the culture, even before the project starts, in how they talk about it and recruit people into the team. They will need to be mindful of what they are saying, because their remarks will be taken to heart by other members of the collaboration.

In 2018, Wellcome published a helpful guide for early career researchers called, 'A Career in Research—Tips for running your own research group' [11] which has some guidance for those embarking on a research career, including sections on leadership and collaboration.

To illustrate how great leadership can enable multidisciplinary research collaborations to flourish, we interviewed the leadership team from the University of Bristol 'Bristol Bridge' collaboration.

3.3.1 Bristol Bridge—a case study in the importance of leadership and shared values

The University of Bristol has a large and active interdisciplinary research community in antimicrobial resistance (AMR). But this was not always the case. Back in 2014, Adrian Mulholland, Professor of Chemistry at the university, wanted to put together a team of researchers to lead a new research network, Bristol Bridge. The team was to apply for a grant from one of the UK funders, the Engineering and Physical Sciences Research Council (EPSRC) to bring together a multidisciplinary community in AMR research across the University.

But his strategy was not the strategy one might expect. He didn't look for the highest profile researchers in each department to join the leadership team, or for those with the highest status. Instead, Adrian looked for authentic people that were interested in collaborating and learning and who were keen to bring their research knowledge and understanding to solving a new challenge.

The people that joined the Bristol Bridge leadership team knew that they would have to put in a lot of effort and energy to build a network that was new and crossed traditional disciplinary boundaries and to put on successful networking events. They were happy to do that because they were contributing to the common challenge of coming up with approaches or solutions to tackle AMR. Reflecting on their experience, another member of the leadership team, Matthew Avison, Professor of Molecular Bacteriology told us, 'It might have been luck [at the time], but if I had to go back and do it again, they would always be the people I would choose.'

When we met with the Bristol Bridge team all of them kept coming back to the importance of the values that underpinned and built the collaboration so successfully. Adrian had assembled a team with the same shared values. It was obvious that these values had been 'lived' by every member of the leadership team. Because of this, the leaders had been fantastic role models. They had demonstrated the culture they wanted to see in the network through their own behaviours.

The values that underpinned Bristol Bridge were:

3.3.1.1 Cherish and encourage diversity
Adrian knew that to work towards solving the massive challenge of AMR he needed a team of people who had an enthusiasm, curiosity, and a desire to work together. He also knew that to generate genuinely new ideas he needed to attract people with new and different perspectives. In short, he not only needed a diverse team of researchers, he also needed a diverse leadership team.

The Bristol Bridge leadership team was diverse scientifically, diverse in background, diverse in gender and diverse in personality type.

3.3.1.2 Success for the project looked like a successful collaboration
For the leaders of Bristol Bridge, it was important that the overall aims of the project were met and not about individual successes. That is not to say there were not individual successes, there were many. But that wasn't what drove the project, or what held the leaders accountable. Every individual in the leadership team became an advocate for the project *as a whole.*

One of leadership team and Associate Professor in Physics, Annela Seddon told us that; '[the] people in the management team were interested in the success of the project *as a whole, rather than the individual successes that it might bring*'. She went on to tell us that, 'The leaders [of the project] advocated for the project rather than for themselves.'

3.3.1.3 Early career researchers are equal partners
The Bristol Bridge leadership team welcomed early career researchers into the network alongside more established researchers. They were really willing to give early career researchers opportunities to play an integral part in the network, and to

be generous with their own expertise and networks. Robert Hughes, now a lecturer in non-destructive testing and then a post-doctoral researcher in Engineering, told us about his experiences.

'Something that was really obvious coming into [Bristol Bridge], was how inclusive everyone was and how quickly and effectively they would wrap you in. Everything you needed to know—[the leaders] would get you trained in the areas that you needed. It felt like there were no boundaries to where you could go, who you could go and talk to, what kind of techniques and tools you could use or learn. And that was, at that stage of career that I was, invaluable. It exposed me to a huge wealth of other things, and other fantastic people.'

'Within a few months I found myself in a biology lab being trained how to stain bacteria. Then I went out to Thailand to collect water samples and apply all of the techniques that we were developing in the lab, to see if I could get them to work in real samples. I found myself in a lab from 8 in the morning to 11 at night. I plated out hundreds of samples onto agar plates that day—I probably plated out more samples in a day than most people do in their PhD—and I thought I was an engineer!'

As well as the Bristol Bridge team, Robert also had strong support from his line manager—the Principal Investigator on the research grant that he was employed on. He told us, 'They were very supportive of expanding your horizons and things like that, they paused the work I was doing at various points so that I could work on these projects.' You can see that his supervisor had values that aligned with Robert's, which meant that it was a good place for him to develop his career.

The experience of Bristol Bridge has also made a material difference to Robert's career. Now a lecturer, he stood out from the other candidates for the position because he had done something a bit different, and had a bit of knowledge of AMR, a bit of knowledge of microbiology and experience in a biology lab. As Matthew Avison, another one of the Bristol Bridge leadership team reflects,

'Now, going forward in his career he's going to be always thinking 'oh, maybe there's something we can do in that area'. As far as I'm concerned, that's a really good outcome—just as much as a £3 million grant [won by someone more established].'

3.3.1.4 There are no stupid questions
A fundamental value of Bristol Bridge, laid out very early on by Adrian Mulholland, was the motto of 'there are no stupid questions'. As one of the team told us,

'It became a bit of a mantra. It was the motto of Bristol Bridge.'

Adrian explained a bit more:

'I tried to ask some fairly simple questions. I tried to make it clear that I would ask quite basic but honest questions—questions that I knew some other people might think were stupid. I learned a lot from this project as a result. So, others would see that while I was the leader of this project, I was prepared to ask simple questions and even, perhaps to make myself look a little bit silly, but I was more interested in the answer.'

As one of the leaders of the collaboration, Adrian was behaving as a role model. He was walking the walk to create a culture where it was OK for people to ask questions and expose that they did not know something. Every member of the leadership team was, at some point, out of their comfort zone. Because people were genuinely curious, wanting to understand and learn, it felt safe for people to ask questions. We saw how important the right kinds of questions can be for collaboration and putting this value right at the heart of Bristol Bridge was one of the keys to its success.

So, the network had great leadership and shared values. Robert Hughes got a lectureship and a new outlook on research. But none of this on its own would satisfy the University's senior managers about the return on investment. How successful was Bristol Bridge when looked at through the lens of cold, hard return on investment?

After two years, the Bristol Bridge network comprised more than 200 researchers across five faculties and two NHS trusts, had funded 17 pump-priming projects and generated more than £3.8 million in follow-on funding. Now, the AMR research community at Bristol is known as 'Bristol AMR'—a cross-faculty research network with £15.4 million of funding supported strategically by the University, and by funders such as UK Research and Innovation, and Wellcome. AMR is also one of the top three priorities for the regional strategic research collaboration GW4, establishing a centre of excellence in AMR research and training in the South-West of the UK. Not bad for a relatively modest EPSRC investment of £600 000.

If you are reading this book, chances are that you are a person that values research collaboration, possibly one that values multidisciplinary research collaboration. That is not to say that single discipline research is bad, or those research leaders that value this more than multidisciplinary research are wrong; however, if your values do not align with those you look to for leadership, it is unlikely that you will feel motivated. The energy you will need to expend to conform in this culture is energy you will not have for your research. You need to look for leaders and colleagues that believe what you believe, especially at transition points, such as putting together a new collaboration.

The Bristol Bridge case study also illustrates the importance of understanding the difference between being a 'Leading researcher' and 'Leading a research team' or being a collaborative leader.

3.4 Leading researcher or collaborative leader?

In its 2018 report, 'Research Culture: embedding inclusive excellence' [12], the Royal Society makes a distinction between the skills associated with being scientific leaders and being

Collaborative Leadership

A collaborative leader sets a new direction by 'orchestrating' collaborative activities drawing in a diverse set of people and following through with a strong hand to get things done'

PURPOSE

INNOVATION
Shaping Strategy
Facilitates idea sharing and knowledge transfer. Embraces change and navigates risk to drive business performance in alignment with the LT vision

INFLUENCE
Communication & Connectedness
Create cross functional collaboration and cohesive relationships to accomplish change initiatives and Performance results. Leverages network and fosters connectedness

PROCESS

PRODUCT

IMPACT
Process and Performance
Leads by example.
Guides process decisively with emphasis on customer experience profitable growth and business values

INSPIRATION
Talent Strategy & Workforce Readiness
Determines right fit talent.
Engages talent from everywhere. Inspires trust and alignment towards a common purpose

PEOPLE

Figure 3.1. Collaborative leadership model [14].

leading scientists. 'Leading scientists are described in terms of individuals pushing the boundaries of research in academia and industry…. Scientific leaders of research groups, programmes and institutions were identified as having a responsibility to advocate for the researchers of the future and develop the talents and skills of their research teams.'

The Collaborative Leader, a concept introduced by Morten Hansen and Hermina Ibarra in 2011 [13], is often used when discussing how best to lead business teams. The model and behaviours easily translate to a research context—research leaders who can build develop and lead successful collaborations. The model is summarised in figure 3.1.

In this model a leader manages or 'orchestrates' four distinct elements of collaboration:
- **Purpose**—'why' the research is being undertaken;
- **Process**—'how' the research will be conducted;
- **Product**— 'what' the outcomes of the research will be;
- **People**—the team that will come together, sharing their 'why' and making the 'how' happen and delivering 'what' will go on to create an impact in the world.

The Collaborative Leader needs to enable what Morten Hansen and Hermina Ibarra call 'disciplined collaboration'. In disciplined collaboration, the business case for collaboration is thoroughly scrutinised, the environment for collaboration is set up and any barriers removed, for example by dealing with behavioural issues and setting up appropriate incentive and reward structures. The Collaborative Leader also takes key decisions to keep the work moving, and to prevent the collaboration getting bogged down in debate and consensus-decision making. These can slow down the collaboration and result in poor decisions and risk averse behaviour. There's a delicate balance to be struck between 'being the one in charge' where necessary and encouraging constructive and open debate. Building trust and ground rules with your collaborators will allow you and them to identify when a leader needs to step in and bring debate to a close as a necessary part of moving forward.

The distinct leadership behaviours identified in the model provide a great blueprint for you to think about your own role as a leader or member of a research collaboration. The key behaviours are:

- connecting people
- attracting diverse talent into your collaboration
- modelling collaboration at the top.

Let's have a look at each of these in turn:

3.4.1 Connecting people

Journalist and best-selling author Malcom Gladwell uses the word 'connector' to describe individuals who have many ties to different social worlds [15]. Award-winning writer Elizabeth Gilbert has a different word to describe these sorts of people, she describes them as 'hummingbirds' [16].

'Hummingbirds spend their lives moving from tree to tree, from flower to flower, from field to field, trying this trying that. Two things happen: they create incredibly rich complex lives for themselves, and they also end up cross-pollinating the world. That is the service that you do if you are a hummingbird person. Because you bring an idea from here to over here where you learn something else and then you take it here to the next thing you do. So that your perspective ends up keeping the entire culture aerated and mixed up and open to the new and fresh.'

Having connectors and hummingbirds in the team is hugely important for any collaboration. They add value to a collaboration by connecting people and ideas, and by curating information and insights for the rest of the team. We saw the importance of networks for serendipity earlier in the book. Connectors, hummingbirds and collaborative leaders can use their positions and networks to connect ideas and resources that might not 'normally' connect or bump into each other. One could say they both create serendipity and capitalise on serendipity.

McKinsey & Company introduced the term, 'T-shaped', to describe individuals with a depth of expertise in their primary area of work and with the capacity to collaborate with others having different areas of expertise. Tom McLeish, Professor of Natural Philosophy at the University of York, describes a 'T-shaped' researcher as having a depth of knowledge in their discipline together with a 'valency'—which enables them to establish a channel of communication with those from other disciplines and understand the issues they are facing [17]. Going back to the analogy of collaborative research as a team sport, Tom McLeish also described the need for a metaphorical 'huddle', where T-shaped researchers can come together and metaphorically put their arms around one another to collaborate (figure 3.2).

For Simeon Yates, Associate Pro-Vice-Chancellor for Research Environment at the University of Liverpool, interdisciplinarity and learning has been at the heart of his career from the very beginning. His doctorate comprised around one-third linguistics, one-third computer and information studies, and one-third social science. He studied for his doctorate at the Open University, where there were no full-time undergraduate students, and spent a lot of his time in the company of other

Figure 3.2. A team huddle is a useful metaphor for a collaboration of T-shaped people. Pixabay: chrisreadingfoto.

researchers talking to people from other disciplines. Simeon really enjoyed that experience, but later in his career he realised that when he became a member of staff it was much harder to spend time working with people from different disciplines than it had been when he was a student. Simeon told us that, 'I've always believed that you learn more and have new ideas if you're talking to people from different disciplines. For me it's part of the fun of being in a university, the unification of knowledge, having the opportunity to talk to different people. For me it's personal, I enjoy it, and I think it's important.'.

From his testimony you can see that for Simeon, being a connector and a learner are key values to him, which have driven his career.

In today's world, it's so important to connect with people outside of your 'normal' circles—this might be business, industry, not-for-profit sector, charity, hospitals, policymakers in government or local authorities. It's worth thinking about how far your networks stretch outside of your 'normal' world. Are you getting the benefits of different perspectives and weak ties to generate new ideas and engineer serendipity into your research endeavours?

3.4.2 Attracting diverse talent into your collaboration

Research has shown that across all sectors diverse teams produce the best results IF they are well led. A collaborative leader both attracts and proactively recruits diverse talent into a team. It's in difference—or diversity—that we often find our most creative ideas and there are also implications for leaders that are worth highlighting here.

When you have a collaboration made up of diverse disciplines the collaborative leader needs to work hard to establish a clear vision and use a vocabulary that cuts through tribalism. Difference can also result in conflict. While creative tension is very good for getting to the best ideas, the collaborative leader needs to hold the space for constructive, creative conflict and not let this degenerate into destructive argument.

We are often asked by academic researchers how an established team can be refreshed and energised to come up with new and different ideas. One answer is to bring in new people! The intellectual stimulus of constructive disagreement will spark original thinking and challenge your assumptions, so bring in people with different talents, backgrounds, and interests to those already in the team, and with skills and knowledge that complement your own.

It is hard work at first, because you need to integrate these people into the team, but it will pay dividends. Work on unconscious bias has shown that static groups can lead to 'group-think' which can stifle innovation and suppress new thinking. Think about the diversity of your team—does it include a mixture of early career and more established researchers? What about the profile of your other partners? Does the age profile reflect the people that are likely to be affected by the research that you are doing?

Egon Zehnder, the recruitment firm, started considering potential, defined as curiosity, insight, engagement, and determination, into candidate assessment. When they did so, they found that the resulting candidate pools were more diverse in terms of race and gender [18]. What factors are you taking into account when you are recruiting people into your collaboration?

'Differences in convictions, cultural values and operating norms inevitably add complexity to collaborative efforts, but they also make them richer, more innovative and more valuable. Getting that value is at the heart of collaborative leadership.' [19]

3.4.3 Modelling collaboration

'Leaders need to model the way' is one of the five practices of exemplary leadership identified in the seminal book, *The Leadership Challenge*, by James Kouzes, Fellow of the Doerr Institute for New Leaders at Rice University, and Barry Posner, Professor of Leadership for the Leavey School of Business at Santa Clara University [20]. They talk about how exemplary leaders set the example by behaving in ways that reflect the shared values and vision of the team. Professor of Business Administration at Harvard University, Francesca Gino, says that leaders who are frustrated by a lack of collaboration can start by asking themselves a simple question: *What have I done to encourage it today?* [21]

As human beings we are highly attuned to detect authenticity (or the absence thereof), so it is vitally important that collaborative leaders are genuine in how they champion collaborative practice. How they behave will influence others much more than what they say. Nothing erodes trust and confidence in a leader more rapidly than being observed behaving in a way that is counter to the team's shared values and vision.

3.5 Power dynamics and their impact on collaboration

Whether a research leader, a leading researcher, or a collaborative leader—these individuals have formal and informal power. In a research collaboration, power differentials and how they are managed is crucial for success. In Bristol Bridge, for example, the power differences between the leadership team and the early career researchers were managed through the shared value that early career researchers are equal partners, and the motto of 'there are no stupid questions'.

The way a leader behaves towards their team and how effective they are as a result can depend on both their values and the source of their power. The same person, in different circumstances, can behave differently because they are drawing on different sources of power.

In research teams, as in many other organisations, there may be individuals who also hold positions of high status in the management structure. People like this probably have responsibilities in the formal governance framework and are held to account for organisational and/or team performance. They could be involved in setting the standards and targets which affect others below them in the career ladder, for example in relation to reward and recognition practice and promotions. This gives them a certain degree of what is called 'positional power'. An example of this kind of position might be a Pro-Vice-Chancellor, or a Head of Faculty or Department.

Their position also enables them to create a culture within their part of the organisation which reflects their own attitudes and values. This will determine how—and if—they share power, and with whom they might share it. They can be the big amplifiers of messages, cultural norms and are seen to be key influencers of 'the way we do things around here'.

Power can also come from a track record of securing funding. Some researchers find that they have power inside and outside of their own workplace which comes from the level or type of funding that they have brought in, their public profile, awards, and prizes, their connections and network and/or being appointed to an influential body. These people can become the institutional or disciplinary 'heroes' or 'superstars'. People with this type of power can sit outside of the traditional hierarchy and can be viewed 'by management' with apprehension—depending on if and how they wield this influence.

How other people respond to power differentials can also affect the dynamics of a meeting or room. Liz remembers leading a workshop with some very influential researchers. In the pre-workshop briefing, she was told, in hushed tones, that one of the researchers who was attending had received funding from the Bill and Melinda Gates Foundation, or 'Gates Funding' for short.

For facilitators this is always an interesting moment, as it immediately alerts us to keep an eye on the impact of the status of that individual on others in the room. It was interesting to see how this played out as the group gathered. In the early stages of the workshop Liz watched other researchers defer to him and his opinions. The trouble was that this was stifling new ideas and constructive discussion. It was time for the facilitators to step in. Liz quietly suggested to the high-status academic that they might usefully use a soapbox—a short uninterrupted broadcast to the room—to share more about his funding with the group.

After the soapbox, everyone settled down. By allowing that person to say what they wanted to say and for others to hear what they wanted to hear, the power differentials were reduced and the senior academic became one of the participants and not a 'guest speaker'. In essence, Liz created a more level playing field, which is essential for creating a psychologically safe place for people to open-up and collaboration to begin.

More on running a soapbox can be found in chapter 2.

Often, the leader of a collaboration needs to behave as a facilitator, aware of differences, alert to any tensions which are bubbling up and to be able to diffuse these constructively, before they become an issue for the team.

3.6 Leaders and followers in collaborations

'... it's long overdue for academics and practitioners to adopt a more expansive view of leadership—one that sees leaders and followers as inseparable, indivisible, and impossible to conceive the one without the other.'—Barbara Kellerman [22]

As we saw in the Bristol Bridge Case Study, the leader gives their followers a clear vision—a reason to act and to collaborate to achieve a common purpose or goal that is bigger than themselves.

Sometimes people find themselves with followers, i.e., as leaders, in positions of high rank and influence, not because they have sought this out but because their ability to inspire others leads to better performance within their teams relative to others. Often these leaders make a connection at an emotional level with those that follow them based on shared values.

In 2020, in during the Covid-19 pandemic, the late Captain Sir Tom Moore, a centenarian, raised more than £32 million for charities that support the National Health Service (NHS) in the UK. People were inspired by him to run, climb, create artworks, and sing to raise money. He started with a dream, to raise £1000 for the NHS during the pandemic as a thank-you for his own treatment. He moved others to donate and to act, not because he set out to be famous, but because he espoused values and a powerful message of gratitude that so many other people were thinking and feeling—'Thank-you, NHS'. His message of hope that, 'tomorrow will be a good day', also resonated in a time of uncertainty and made others feel that they too could make a difference, no matter how old, how infirm, or how young.

If you have worked with a leader that has empowered you to achieve, you may not remember each and every meeting or conversation, but you will remember how that person made you feel. The experience is deeply personal and lasting. When we believe in the vision created by a great leader we will go above and beyond for the good of the team or the cause. The converse is also true of course. It is unlikely that you will want to follow leaders who make you feel belittled or taken for granted, or where you don't share their vision and values. If you have to follow them, because of their positional power for example, it is likely you will do so grudgingly and will likely be much less effective as a result. And the effects can go much further than your performance at work. Working for a poor leader can increase your level of stress and be bad for your health, with studies indicating a link between work-related stress and coronary heart disease [23].

3.7 The i-sense-AHRI collaboration and the importance of leadership

The fight against HIV in South Africa is the key goal for a collaborative partnership between the UK's Engineering and Physical Sciences Research Council (EPSRC)-funded Interdisciplinary Research Collaboration in Sensing Systems (i-sense) and the Wellcome-funded Africa Health Research Institute (AHRI) in South Africa. Their Medical Research Council (MRC) funded 'm-Africa' Global Challenge Research Fund collaboration uses mobile technologies to test and treat those hardest hit by HIV.

In 2016, this collaboration was co-led by Rachel McKendry, Professor Biomedical Nanoscience at University College London and Deenan Pillay, then Director of AHRI and Professor of Virology at UCL. Rachel and Deenan knew that this partnership was crucial for the success of a collaborative research project, which explored whether key smartphone functions, such as the camera, can interpret HIV test results and securely send them to local clinics, supporting virtual follow up appointments and rapid treatment. You can read more about this fascinating project on the i-sense website [24].

For this project to be a success, there needed to be a new collaboration built between colleagues at AHRI, the clinicians in KwaZulu-Natal and the team of researchers from i-sense. Rachel and Deenan led by example, presenting the whole i-sense collaboration to AHRI, starting to build the relationships of trust which would sustain the collaboration, and to engender some excitement and enthusiasm around the common goals of the two organisations. When we spoke to some of the postdocs that had worked on the project, Michael Thomas, Eleanor Gray and Valerian Turbé, they told us how vital this leadership had been to the future success of the project. When the postdocs arrived in South Africa they found a very warm and welcoming environment, where people were keen to see them succeed. Before they had left the UK, Rachel had made it very clear to the postdocs and students that they had a crucial role in strengthening and maintaining the relationship that had been created, by involving people, communicating, and feeding back to people at all levels in AHRI how their contributions were moving the project forward.

It was clear, when we spoke to the postdocs, that they had really taken this message to heart and 'gone the extra mile' to make a positive impact and involve people in their research. Mike and Valerian told us how important that had been to them, and about the positive impacts that also had for the research; that it inclines people to help you and to prioritise your project among other demands on their time. They spoke passionately about the huge benefits that this had had for the postdocs and for people in AHRI—that it 'works both ways'. The shared values and vision permeated throughout AHRI and i-sense from top to bottom, a great example of successful collaborative leadership in practice.

3.8 The 'soft stuff'—psychological safety and why it's important for collaboration

We have already talked a lot about the fact that collaboration is a human experience and that there are always human dynamics in collaboration. It is often said that working collaboratively is an integral part of academic life, yet we know that in practice it can be much more difficult than expected.

When working with colleagues do you feel comfortable to ask questions, disagree, highlight errors or suggest whacky ideas? The answer to that question will depend on the colleagues in question and whether the meeting or workshop that you are in is psychologically safe.

The leading global researcher in psychological safety, Amy Edmondson, defines psychological safety as [25]:

'The perception of the consequences of taking personal risks in a particular context… [and]…a shared belief that the environment is safe for personal risk-taking.'

A psychologically safe environment is one where it is OK to be both candid and vulnerable and where the values of the group support this. It's an environment where it is OK to break down the 'walls' that we put up to make us feel safe. Walls, like discipline silos, within which there are cultural norms, and you are part of the tribe.

When we are trying to get into collaboration with people we don't know well, we may perceive there to be threats to our status, autonomy, sense of relationship and sense of fairness. We've seen already how a scarcity mindset has a highly destructive effect on our ability to collaborate and share. Scarce resources, or the perception of scarcity, competition, and uncertainty about the task are also psychological threats.

'Uncertainty and the prospect of failure can be very scary noises in the shadows'
—Tim Ferris

Where there are threats, there is fear, and fear gets in the way of creativity, inspiration, and productivity. Feeling fearful makes us pull back cognitively—fight or flight in action—and our brain shuts down our higher functioning and thinking in preference to our survival. Fear actually shrinks the size of our hippocampus, the part of the brain that is critical for long-term memory, and which plays a key role in creativity and imagination. Fear prioritises the short term and the easy, leaving the long-term and harder challenges for another day. It makes us prioritise activities and actions that reduce the threat, not those that support a collaboration. It's obviously important for collaborations to succeed that we dial down the fear and threats, and that means making them psychologically safe.

'Named your fear must be before banish it you can'—Yoda, Empire Strikes Back

In a psychologically safe environment people can challenge deeply held beliefs and start navigating through the landscape of rational and emotional responses to working with others. People will sense whether or not they will be OK if they experiment with the uncertainty and tension that can come with collaboration. That's not to say it's comfortable—there is likely to be plenty of uncertainty and disagreement, and a feeling of edginess, but there is also likely to be support, curiosity and energy. It's about forming a sense of community where people feel that they will be treated sympathetically and giving people the opportunity to have their comfort zones respected, appreciated, and gradually stretched.

We saw earlier in this section how important diversity, in all its dimensions, is for creativity and innovation. Leaders need to encourage the people they work with to feel safe enough to reveal and apply their different ways of thinking. There needs to be time and the space to experiment and try things multiple ways. A strong sense of psychological safety will lead to better questions and better research outcomes as a result. Alison Reynolds and David Lewis describe cognitively diverse and psychologically safe teams as [26]: Curious, Encouraging, Experimental, Inquiring, Nurturing and Forceful. They clarify that forceful in this context was forceful in solving problems, NOT with each other!

Different sizes of teams and types of work require different degrees of psychological safety—large, fluid groupings and distributed teams working on complex and changing problems have different needs and norms compared to a small, stable group working on a tightly defined task. In general, though, psychologically safe environments have a focus on the problem or challenge, not on individuals.

'The secret is to gang up on the problem, rather than each other.'—Thomas Stallkamp, leading US industrialist

3.8.1 How do I know if I am in a psychologically safe space, or not?

It's likely that instinctively you know whether your team, meeting or event is psychologically safe—human beings have become highly attuned to these signals over millennia. Have a read through the list of 'key features' and 'red flags' in the box and relate them back to your own experiences where collaborations and teams have worked well, or not. Not all collaborations are perfect, and may have a mixture of positive and negative characteristics, but the ones that work have, on balance, more of the positive signs of psychological safety than the red flags. It's possible to improve how an existing team or collaboration works by tipping the balance towards the safe environment and dealing with the warning signs.

To bring diverse groups together to collaborate successfully we need individuals to be able to admit ignorance and uncertainty, to voice opinions and concerns and to work with others who are different to ourselves. Facing these interpersonal threats is inherently risky and it becomes even more so when there is a differential in the perceived levels of expertise, power or status. If we don't deal with these issues and create a safe environment it's likely that a collaborating team won't be able to perform well.

Key features of a psychologically safe space

Curiosity, listening to understand, open to new ways of thinking
Divergent thinking
Displays of vulnerability, admitting mistakes collegiate, no blame
Speaking up about concerns
Generosity, sharing information and knowledge freely, sharing suggestions for improvements/new ideas
Proactive and abundant mindsets
Gratitude
Integrating perspectives and ideas, building on ideas of others
Feelings of vitality-energy
Creative
Risk-taking
Problem solving
Everyone contributing equal amounts of air time
Trusting relationships
Embraces diversity, celebrates difference, and celebrates commonality.
Autonomy
Open and reflective questions
Deep listening

Warning signs (red flags) for psychologically unsafe spaces

Anxiety
Power games, high-status individuals using hierarchy—'I am better than you'—power imbalances, leaving people out
Elephants in the room, not reporting errors/issues
Attributing blame, blame culture, sacrificing people, 'throwing people under a bus',
Defensive and scarcity mindsets
Avoiding interpersonal risk, people feeling inhibited in contributing
Interruptions, tolerating disrespect
Cosmetic listening
Neuroticism
Task uncertainty, Going 'off task', Diversionary /irrelevant questions, comments or discussions
Absence of trust
Absence of relationship
Resentment
Ambivalence

3.9 The importance of 'belonging'

In her 2019 Netflix film, *The Call to Courage*, professor, lecturer and podcast host Brené Brown describes how a group of schoolchildren helped define the difference between belonging and fitting in. 'If I get to be me, I *belong*,' one told her. 'If I have to be like you, I *fit in*.' In a collaboration, belonging means being able to bring our authentic selves to the team, and to feel appreciated and confident for being who we are.

Belonging is inherently inclusive and is an approach to equality and diversity issues that encompasses both majority and minority groups and embraces intersectionality. It's for this reason that belonging is being hailed as the evolution needed to progress equality, diversity, and inclusion at work. We need to attract and retain researchers and innovators into our collaborations whatever their background, age, and experience—creating a sense of belonging can help us to do this.

We all have our 'in-groups'—people that we share particular qualities and values with; and our 'out-groups'—people that we don't. Teams and organisations where people feel a strong sense of belonging have bigger and more diverse 'in-groups' where people are not wasting time and energy trying to fit in. They are able to address more complex problems and take more objective decisions with fewer unconscious biases.

If we can create a sense of belonging in our collaborations, it is likely that we will see better outcomes from the research, because belonging also increases performance. People who feel a strong sense of belonging in their collaboration are more productive, more likely to contribute at their full potential, more motivated and committed and more creative.

Whether we lead or are a member of a particular team we need to ask ourselves—does everyone in this team feel a sense of belonging? How do I behave around this team so that others know they belong? How to I make it safe for people in this team to bring their whole selves to work?

To optimise our research and innovation collaborations we need to create strong feelings of belonging to our professional relationships, to our teams, research groups, collaborations, departments, businesses, and networks. This is particularly important where the power differentials and cultures are very different; for example, when working with groups representing those living in areas of high deprivation. Creating a safe environment, respecting all forms of knowledge, listening to understand and being curious are vital behaviours to encourage a sense of belonging and commitment to the work of the team.

3.10 How to go about creating a psychologically safe atmosphere?

During collaboration, particularly when initiating new relationships and/or in a large and diverse group, it can be helpful to structure conversations to enable everybody to feel safer. The absence of structure can allow more extrovert and/or high-status individuals to dominate. Someone who isn't speaking up may feel unsafe, but don't assume so. It may be that that person isn't yet ready to contribute and is formulating their thoughts before wanting to speak. Some people think through talking their way to a conclusion. The benefit of a well-structured

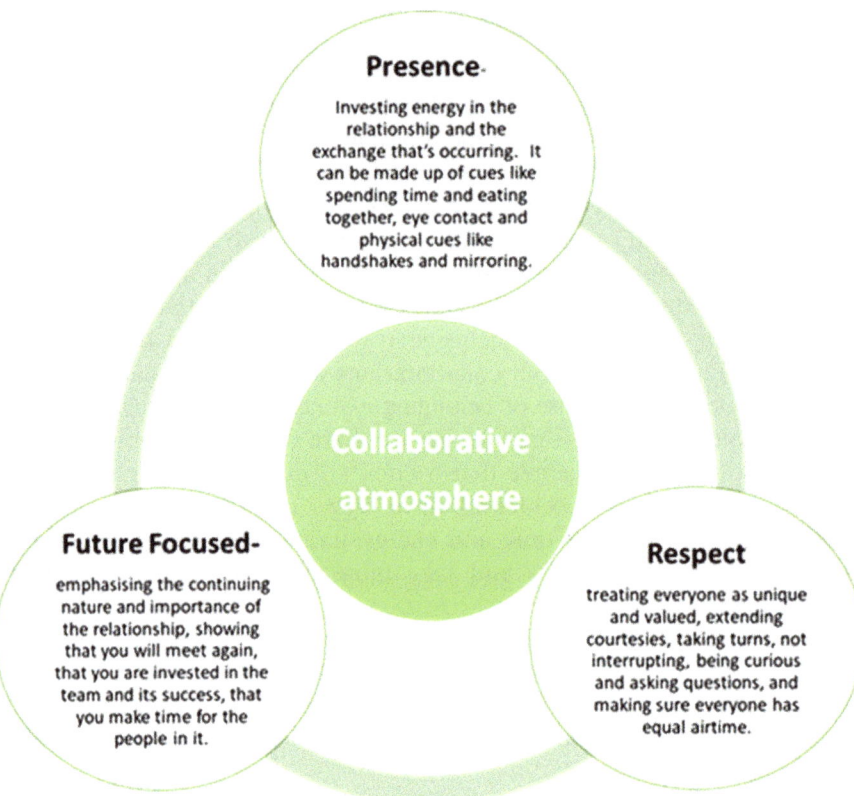

Figure 3.3. Belonging cues, after Coyle [27].

conversation empowers individuals with different ways of thinking to contribute freely within a framework that steers the discussion in a neutral and balanced way. It provides everyone with an opportunity to have the space to react and interact without any individual being exposed or taking over.

Author Daniel Coyle in his book, *The Culture Code* [27], highlights the importance of 'belonging cues'—a language of belonging which matters more than words. Belonging cues are made up of small signals repeated again and again and again, signals like spending time together, listening, taking turns, mirroring, and eye contact. He summarises them into three basic groups (figure 3.3).

By showing that we are interested in what our colleagues have to say, that we care about them, that we are curious about their perspectives and experiences, we can create a sense of belonging. Leaders that create a culture of belonging reinforce that everyone is important to our research and innovation system, that we all have a role to play and will be encouraged to play that role to the best of our ability. Hearing something from a leader will not change the behaviour of others; it is seeing the leader communicate the behaviours through their own actions that makes the difference in the language of belonging.

By asking for feedback and for help, and by not allocating blame or 'throwing colleagues under a bus' to spare our own blushes we can create trust and psychological safety. By reinforcing that it doesn't matter what school you went to, or what region your accent is from, or whether you have a doctorate we can allow more ideas to come forward.

We think that it is possible to create a culture of belonging in diverse multi-disciplinary teams. We believe that by adopting simple behaviours we can help everyone involved in research and innovation feel that they belong. All of us need to help create this culture; led and role-modelled by collaborative leaders and those in positions of real influence.

3.11 Key learning points—chapter 3

- Collaborative leadership is essential for research collaboration.

- Collaborative leaders connect people, attract diverse talent into their team and model collaboration at the top.

- Diversity in collaboration takes many forms and is important for avoiding group-think.

- Diversity can introduce friction which can be channelled usefully to create new ideas.

- Research collaborations take many forms and involves lots of different types of roles.

- The best collaborative leaders are not necessarily the world-leading researchers.

- Being a collaborative leader is a service you give to your team. It demands a range of interpersonal skills and a significant time commitment.

- Values underpin collaborations; it is the responsibility of collaborative leaders to communicate the values and behaviours of the collaboration through their own actions.

- Leaders require followers; people will wholeheartedly follow those whose vision and values align with their own.

- A sense of belonging to your collaboration can increase performance, productivity, motivation, commitment and creativity.

- Psychological safety reduces the perceived level of threat of the situation. It supports being curious, stretching our comfort zones, and challenging our deeply held beliefs and opinions.

- Structured conversations support psychological safety and belonging.

- Belonging cues are small but important signals and rituals repeated over-and-over again.

References

[1] https://research.fsu.edu/research-offices/ord/collaborative-collision/upcoming-events/
[2] https://huffpost.com/entry/collaboration-vs-cooperat_b_10324418
[3] Syed M 2019 *Rebel Ideas—The Power of Diverse Thinking* (London: John Murray Press)
[4] Snow S 2018 *Working Together Without Falling Apart* (Penguin)
[5] Reynolds A and Lewis D https://hbr.org/2017/03/teams-solve-problems-faster-when-theyre-more-cognitively-diverse
[6] Hansen M T and Ibarra H 2011 *Harv. Bus. Rev* https://hbr.org/2011/05/getting-collaboration-right
[7] https://youtube.com/watch?v=5IXAO-YtwjA
[8] https://youtube.com/watch?v=S1pcwVAdmwk
[9] Sinek S 2019 *The Infinite Game: How Great Businesses Achieve Long-Lasting Success.* (London: Portfolio/Penguin)
[10] Wambach A 2019 *WOLFPACK: How to Come Together, Unleash Our Power and Change the Game* (New York: Celadon Books)
[11] https://wellcome.ac.uk/sites/default/files/research-careers-tips-running-esearch-group-2018-05-17.pdf
[12] https://royalsociety.org/-/media/policy/Publications/2018/research-culture-workshop-report.pdf
[13] Hansen M T and Ibarra H July 2011 *Harv. Bus. Rev.* https://hbr.org/2011/05/getting-collaboration-right
[14] After Estis & Associates https://slideshare.net/restis/ryan-estis-associates-collaborative-leadership-model
[15] Gladwell M 2000 *The Tipping Point: How Little Things Can Make a Big Difference.* (New York: Little, Brown and Company)
[16] https://listennotes.com/podcasts/oprahs-supersoul/elizabeth-gilbert-the-Rpr3EiYbAmC/
[17] University of Durham 2020 *Knowledge Across Borders Webinar Creating Knowledge Across Disciplinary Boundaries* https://youtube.com/watch?v=gPAUMdrwxVE&feature=emb_logo
[18] Gino F and Staarts B November 2015 'Why organisations don't learn' *Harv. Bus. Rev.* 112–18
[19] Ibarra H and Hansen M T 2011 Are you a collaborative leader? *Harv. Bus. Rev.* 68–74 (July–August)
[20] Kouzes J and Posner B 1995 *The Leadership Challenge* 2nd edn (San Fransisco, CA: Jossey-Bass)
[21] Gino F 2019 Cracking the code of sustained collaboration *Harv. Bus. Rev.* 73–81 (November–December)
[22] Kellerman B 2007 *Harv. Bus. Rev.* **85** 84–91
[23] Sara J D, Prasad D, Eleid M F, Zhang M, Jay Widmer R and Lerman A 2018 Association between work-related stress and coronary heart disease: a review of prospective studies through the job strain, effort-reward balance, and organizational justice models *J. Am. Heart Assoc. Cardiovasc. Cerebrovas. Disease.* **7** e008073
[24] https://i-sense.org.uk/news/using-mobile-technologies-test-and-treat-those-hardest-hit-hiv
[25] Edmondson A 1999 Psychological safety and learning behavior in work teams *Admin. Sci. Q.* **44** 350–83
[26] Reynolds A and Lewis D https://hbr.org/2018/04/the-two-traits-of-the-best-problem-solving-teams
[27] Coyle D 2018 *The Culture Code: The Secrets of Highly Successful Groups.* (New York: Bantam Books)

Chapter 4

Creating a collaborative organisation

'A truly multidisciplinary environment is really a philosophy. The institutional culture has a huge part to play and can make or break whether individuals can collaborate successfully.'—Sebastien Ourselin, Head of School, School of Biomedical Engineering & Imaging Sciences, King's College London.

The Wellcome Trust report, 'What researchers think about the culture they work in' [1] talks about the importance of collaboration and creating a culture where collaboration is encouraged and celebrated. When asked to provide three words that described an ideal research culture, respondents most commonly said supportive (20%), collaborative (17%) and creative (6%) (figure 4.1).

So far, we have explored how to prepare yourself to collaborate, and how to develop a collaborative team. Institutions and organisations also need to play their part in rewarding collaboration and putting collaborative activities at the heart of their culture. In this chapter we are going to look at the organisational context. How can organisations create the conditions for collaborative behaviours to become 'the way we do things around here' [2]? To do this, we are going to be using a systems approach to think about how an organisation can move towards becoming truly collaborative.

Structures and organisational norms can get in the way of collaboration—from the way people are incentivised to the use of buildings and the way internal budgets are managed. Transformational leaders are thinking about how they can orient the whole of their organisations behind new ways of working and across disciplinary boundaries. They are walking the walk of collaborative leadership and showing people in their organisation that collaboration really matters.

Elements of the previous chapters will also apply to an organisation. A team is, in itself, a system. Here we will concentrate on those aspects that emerge in an organisation which is made up of multiple team-systems. This could be a department, a faculty or an entire university, business or charity. The interactions between

Figure 4.1. Words that researchers would use to describe an ideal research culture [1]. (Wellcome Trust CC BY 2.0.)

the different parts of the organisational system will give rise to features which are not always easy to anticipate or control, and yet which impact, both positively and negatively, on the ability of people to collaborate.

4.1 Organisations as complex systems

'*A system is an interconnected set of elements that is coherently organised in a way that achieve something.*'—Donella Meadows [3]

The work of Siv Vangen, Professor of Collaborative Leadership at the Open University Business School, highlights the complex and systemic nature of collaborations:

'*…collaborative contexts are complex webs of overlapping dynamic hierarchies and systems that comprise competing designs and processes that are necessary to achieve the desired outcomes.*' [4]

Organisations vary hugely in size but are all made up of people who have come together to achieve a common purpose. The combination of individuals, teams, behaviours, structures, skills, technologies, and processes that all together make up the organisation also combine to form a 'complex system'. There is a great deal of research and literature available, both about complex systems and about systems thinking applied to organisations, and we will not attempt to cover this in any depth here. Interested readers who want to dig in further might like to start with the work of Donella Meadows or Peter Senge [5] in systems thinking and organisational learning. Instead, we will just summarise some of the key concepts and focus in more detail on their implications for creating a 'collaborative' research organisation. We'll explore the 'Johnson and Scholes' model of the Cultural Web to analyse and describe collaborative cultures. We will use case studies to illustrate where the model can help us to identify barriers and enablers for research collaboration within and between organisations.

In an organisation, the 'system' is the combination of the people, processes, structures, technologies, skills, and behaviours that work together and enable the organisation to deliver. These might also be thought of as organisational habits [6]. Through understanding connections across the organisation, we can come to better understand how to solve problems and develop new habits.

Organisations are also 'complex systems', which means that the various elements of the system interact with each other in ways which are not entirely predictable and give rise to what is called 'emergent behaviour'. Small interventions can have very large impacts, and what might be expected to have a large impact can lead to little or no measurable outcome. Feedback plays a large role in the complex system of an organisation. An organisation can be thought of as a living ecosystem that uses feedback to self-correct or to move forward. When we hear that people want to 'fix' systems, we know that there may be trouble ahead. As Donella Meadows explains:

'We can't impose our will on a system. We can listen to what the system tells us and discover how its properties and our values can work together to bring forth something much better than could ever be produced by our will alone.' [3]

The broader system within which the organisation operates, which some might see as an 'external' system, can exert a huge influence on an organisation and its attitude towards collaboration. It affects the organisational leadership and the people working there.

In turn, how individuals view collaboration can impact on internal and external partnerships and by so doing have an impact on the collaborative performance of the whole system (figure 4.2).

In an ideal world, this would form a virtuous upward spiral of increasingly effective practice. In the real world, systems are complex. Emergent behaviours mean that progression towards a desired goal may take a 'two steps forward, one step backwards' type of pathway!

The broader external system (which is also a system comprised of systems) will include other organisations such as research funders, government, businesses, and charities, to name just a few.

It's the complexity of this innovation system which has made multidisciplinary collaboration the subject of research and innovation strategies for decades. It has also made the so-called 'valley of death' so resistant to initiatives and structures designed to bridge the gap between innovation in our universities and commercialisation by businesses.

Thinking about an organisation holistically as a system allows recurring patterns, trends, feedback loops, and cycles to be identified which might not be found if you think of it simply as the sum of the actions of individuals and teams. Applying systems thinking to an organisation is not about one-off issues or short-term outcomes. It is a long-term process to identify the consequences of decisions on other parts of the system.

Because organisations make huge numbers of decisions, and have emergent properties, there is no feasible way to predict all the long-term outcomes of all the decisions

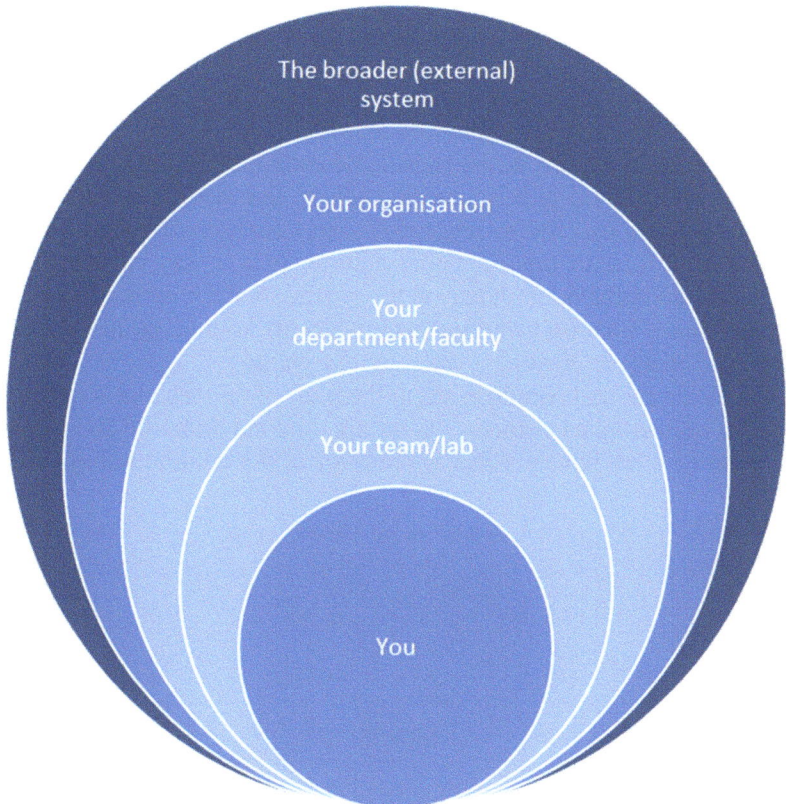

Figure 4.2. Schematic representation of nested systems relevant to an individual researcher.

made. Nonetheless, by reviewing the organisation holistically *some* of the unintended, and usually unwanted, consequences can be identified. Once recognised, it becomes possible to take action to mitigate undesirable impacts, or systems failures. It is also possible to amplify any emergent benefits to improve organisational performance.

By taking a holistic or systemic view, so-called 'leverage points' can also be pinpointed —the situations where making an intervention can lead to the most meaningful and desirable change outcomes. It can be useful to reflect on your own organisation's 'leverage points' for collaboration, as this can help to make a significant difference to:
- how individuals and teams within the organisation interact;
- how ways of working and processes affect one another;
- how easy or difficult it is for individuals and teams to collaborate with each other and with other organisations.

In their white paper for the University of North Carolina, executive coach Kip Kelly and rock star turned entrepreneur Alan Schaefer describe how, when collaboration is focused only on teams or a single level within an organisation, it is extremely difficult to sustain, and this makes the benefits of collaboration fleeting [1]. They say that organisations must redefine what collaboration means by making it part and parcel of the organisational culture and daily operations. This is where the cultural web model comes in as a useful tool for thinking about organisational collaboration.

4.2 Creating a collaborative organisation using the cultural web model

Creating a collaborative organisation means creating an organisation that has a culture of collaboration both internally, across teams, and externally with people from another organisation(s). The cultural web is a useful tool developed by two academics in the fields of business, leadership and management, Gerry Johnson and Kevan Scholes [7], which can be used to map organisational culture. It is a way of representing the influences, values, and behaviours at play in the organisational ecosystem. It can be used to identify what is and what isn't working, and to identify leverage points where you can really make a difference in creating a collaborative culture.

In this model, the cultural paradigm sits at the heart of the web and is surrounded by six overlapping or interlinked elements. As we are thinking primarily about the culture of collaboration here, we have put this at the heart of our cultural web (figure 4.3).

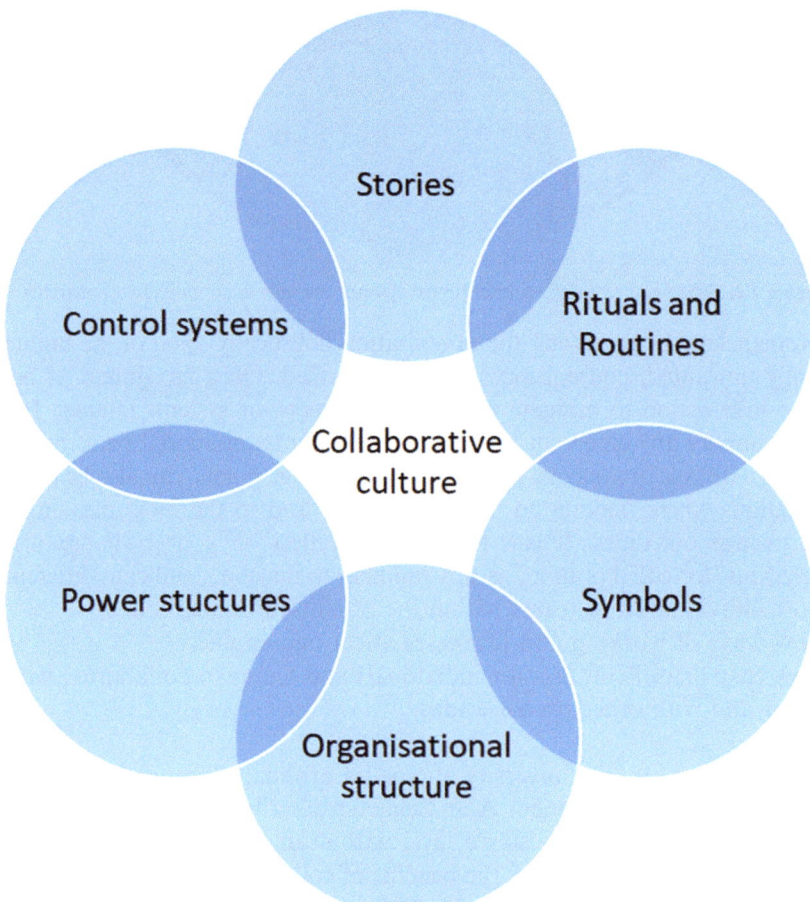

Figure 4.3. Schematic of the cultural web for a collaborative cultural paradigm.

4.2.1 Stories

Stories are how an organisation explains itself and its history. They reflect what is valued, who the heroes are and what desirable outcomes look like.

- How do people within the organisation talk about where they work, especially when new people join?
- Do the stories that you tell reflect a workplace with a collaborative culture?
- What messages do you and your team communicate during formal inductions?
- Do onboarding frameworks include a broad range of departments and functions or are they narrowly focussed around job roles and families?
- What are the stories that you tell about instances of 'failure'?

Postdoctoral researchers Michael Thomas, Valerian Turbé and Eleanor Gray told us about their experiences in South Africa visiting the Wellcome-funded Africa Health Research Institute (AHRI) as part of their work with from the Interdisciplinary Research Collaboration in Sensing Systems (i-sense), funded by the Engineering and Physical Sciences Research Council in the UK. They gave us a great example of the use of stories to ensure that all the people working in AHRI were invested in the success of their collaboration:

'Someone working in IT support isn't just in IT support, they would tell you that 'I am ensuring that the facilities are there to enable research to happen'. The building and maintenance team were really excited about our collaboration—that just never happens! This works well because no-one sees their job as just a source of frustration, or a wage, they see it as bigger than that. They feel that they are actually part of the research team, facilitating the research, or doing different parts of the research. That was really interesting and not something that I've seen in other places that I've worked.'

This is a great story about the type of place AHRI is to work. Everyone knows why they are there and that they are contributing to something bigger than their own role, something that gives their 'job' meaning. Mike, Eleanor, and Valerian also reflected on the relative uniqueness of this sense of belonging, commenting that it was not something that they had seen as explicitly in other organisations.

Anthropologists have long known that stories are an important way for human beings to make meaning of the world around us. Storytelling has been found to be a powerful way of establishing and maintaining social norms and cooperation [8], and by extension, for creating an organisational culture. If we want a collaborative culture, we need to tell stories about our people and our organisations that can inspire others and motivate them to act in a collaborative way.

Mary Barra, CEO of General Motors, in an interview for the HBR Ideacast [10], described how important stories are for the culture of her organisation, and how they link to senior leaders' behaviours and reward and recognition processes [9]. She said:

'Do people leave work and say, "I have a great day. I'm working for a company that's working toward a very important vision or mission or purpose and I feel good about it," or do they go home every day and they're frustrated?'

'Because the culture is the stories people tell about what it's like to work there. So, as we started to hold each other as senior leaders accountable to live those behaviours every day and consistent with our values, that started to change our culture.'

'Then we rolled them out to every employee in the company. Our compensation system is linked to demonstrating those behaviours. Our recognition system is linked to say, "I want to recognise somebody for being bold," or looking over the horizon, or thinking customer.'

'It's really allowed us to, I think, start to make a very important cultural transformation. We're not there yet. I don't know if we'll ever be there, but we know what we're moving to. It's clearly communicated and it gets reinforced every day.'

There are many, many books, research papers, blogs, and articles that explore how and why you might choose to use storytelling as a communications tool for business and innovation [10]. We would encourage you to reflect on the stories people in your organisation tell each other about collaboration, those that are told to people outside of the organisation and even the stories that you are telling yourself. Because storytelling is such a powerful way of communicating, if you are telling stories that don't convey the messages you are hoping to get across, you are unlikely to get the behaviours, practices, and outcomes you are looking for.

4.2.2 Rituals and routines

Rituals and routines are the accepted ways of working and what people expect to happen in any given situation. Sometimes these get formalised into policies, which then become part of the organisation's control system.

Rituals have been studied for many years by cultural anthropologists and are known to calm the mind and reduce anxiety. When Annette played competitive badminton, she had a particular ritual before serving which allowed her to regain her breath from the previous point, settle her nerves and become present. It was a small thing, taking only a few seconds, but if she rushed this, or was interrupted, the serve would end up in the net. You can see these sorts of rituals in any tennis match, from amateur to professional. Rafael Nadal has a number of rituals and routines, which some of his opponents find off-putting. Nadal says that 'When I do these things it means I am focused, I am competing—it's something I don't need to do but when I do it, it means I'm focused' [11].

In one academic study which led to the fabulously titled paper 'Don't Stop Believing: Rituals Improve Performance by Decreasing Anxiety' [12] performing a ritual before singing a well-known song, karaoke-style, allowed people to perform better by reducing their anxiety. A great blog post by journalist Katie Morell for beautiful.ai goes into more detail about 'How Rituals at Work Boost Team Performance' [13] and provides more information for anyone who wants to dive deeper into this fascinating topic.

In chapter 3 we covered how important psychological safety is for collaboration and why. Rituals can be used to help create a safe space for new relationships and collaborations to form. Unfortunately, rituals can also be used to exclude, or to inhibit change to the status quo. You may be able to think of a time when you joined a new team and did not really feel that you belonged until you had participated in the routines of that group. This might include, for example, a practice of taking it in turns to bring cakes for the team, or perhaps going to a restaurant to eat together to celebrate an event or milestone in a project.

It's not hard to see how rituals and routines can be used to exclude others, accidentally or deliberately. Late afternoon drinks receptions involving alcohol are a good example of an event that could exclude lots of different people. If you question the purpose of a particular practice, you can sometimes be told, 'we always do it like this', or 'we tried something else, but it didn't work'. This may be a polite way of telling someone to abide by the established routine of the group, or not to rock the boat.

In your team, department, and organisation there will be rituals and routines and you can begin to think about the extent to which these rituals help or hinder collaboration.

- To what extent do your rituals reflect values and beliefs associated with collaboration?
- Are established rituals supporting or actively suppressing collaboration?
- What messages do you communicate during training, deliberately or accidentally?
- How can you use your rituals to celebrate and encourage collaboration?
- How do your rituals include or exclude others?
- Can you use routines and rituals to create a sense of safety and belonging, particularly in new or emergent collaborations?
- How open are you to modifying your routines and rituals?

4.2.3 Symbols

Symbols are the official and unofficial representations of the culture. These can include physical symbols, like logos, jargon, t-shirts, lanyards, mugs, lapel pins, dress codes and how spaces are arranged, or offices are allocated. The Graphene Engineering Innovation Centre (GEIC) at the University of Manchester has a name which describes exactly what you would expect to find within it and the name is the first symbol that you encounter when you approach the building [14]. The University of Manchester is world renowned for its expertise in Graphene, and it's vitally important to ensure that this is communicated consistently.

This part of the cultural web also includes the image associated with your organisation, from both internal and external perspectives. Earlier in the book we talked about the importance of language for collaboration. You can think about the extent to which specialist language is a feature of your organisation or department and what you might do to ensure that potential collaborators are not excluded

 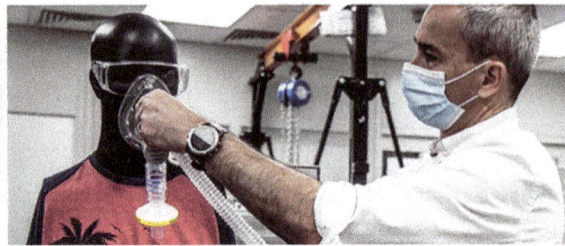

Figure 4.4. UCL Ventura team and the t-shirts that became a symbol for them [15, 16] Credit: ThisisJude (www.thisisjude.uk).

accidentally. Something simple like ensuring visitors know your organisation's dress code can allow them to make choices and help them to feel more comfortable in an unfamiliar space. For example, skirts and clean room suits are not the most practical of combinations, (speaking from experience), and knowing that we might need to don clean room suits would certainly influence our choice of apparel!

There is a crucial difference between symbols that are part of an organisational brand which might be imposed or mandated, and one that a collaborative team embraces for itself because the symbol brings meaning to, and unites, its members. The University College London (UCL) Ventura team, that we met in the first chapter of this book, found themselves adopting an unconventional symbol when they were holed up together in central London in the early days of their project, which we described right at the start of the book.

Tim Baker told us about the eerie nature of experiencing London going into lockdown while the team of engineers kept on working, night and day, to reverse engineer the CPAP device needed by hospitals to keep Covid-19 patients out of intensive care (figure 4.4).

'As London was shutting down the engineers were left to fend for themselves. Meals had to be ordered using a newly acquired UCL Deliveroo account. The team had come together direct from UCL and Mercedes AMG High Performance Powertrain, and it was about 3 days in when they realised that they were going to need to get hold of some fresh clothes. They sent one of the team into Kings Cross St Pancras station, where a few shops were still open, to see what they could find. Much to the team's astonishment, they returned with a bundle of pink t-shirts with Buena Ventura logos on them. It was all that was available—and so you had the team rolling around the centre of London in pink t-shirts, looking for all the world like some stag-do that had gone horribly wrong!!'

The Buena Ventura t-shirts acted like a voluntary dress code for the team. While these t-shirts would not have been ones that the team would have ordinarily chosen to wear (indeed, Tim told us that the intensive care consultants were grateful *not* to have to wear them!), these were not ordinary circumstances. They had already started to come together as a team working towards an extraordinary goal, and the t-shirts cemented the sense of 'all being in it together'.

Not only that, but the t-shirts inspired the name for the product, UCL Ventura. You can see a mannequin wearing one of the t-shirts and demonstrating the device in one of the publicity photographs. Tim told us that there was some resistance from outside the team to putting a name like that on the device, but he understood the significance of the t-shirts and the almost comical 'in-joke' that had brought the team together against the odds. He knew that it would mean a lot to the people involved to have the Ventura name on the device and convinced the sceptics that it would be worth it. Now the UCL Ventura device has gone global, and many of those whose lives are saved by it will not know the story of the device or the significance of the name. For the engineers that worked tirelessly to innovate in truly challenging circumstances, the name and the t-shirts are true symbols of the spirit and commitment of the team to the goal and to each other.

4.2.4 Organisational structure

This is the formal structure of functional units and hierarchies defined by the organogram. Where are the formal lines of reporting and/or authority? Where are the informal lines of influence and who are the influencers? How can these be configured most effectively to support collaboration?

The technology company, Apple, is deliberately organised to support collaboration [17]. It has a functional structure, which evolves in response to organisational growth and technology developments. They have hundreds of specialist teams, dozens of which may be needed for just one component of a new offering. Because no single function is responsible for a product innovation, collaboration is essential. For people to reach and remain in a leadership position, they must be extremely effective collaborators and willing to engage in collaborative debate with other team leaders.

4.2.5 Power structure

The power structure is the set of people with real power to get things done. These are sometimes, but often not, the people in the higher echelons of the organisation. These people have the most influence on decisions, operations, and strategy. What do they really believe and value? Do they champion collaboration? It's an interesting exercise to reflect on your own department or organisation and think about where real power sits. Some people have power to get things done, and others have the power to resist things getting done.

Studies on these 'organisational influencers' have shown that often these do not sit high up on the organogram [16]. Organisational influencers tend to have strong social networks and professional credibility. Do you know who they are? Are your organisational influencers collaborative by nature? Do they encourage or resist collaboration building? You may need to invest particular effort in ensuring that your influencers are engaged and understand what you and the organisation is trying to achieve through collaboration. They may be able to provide you with useful feedback and intelligence, so listen to what they have to say—and remember to thank them for their commitment and support.

4.2.6 Control systems

The control systems are the set of checks and balances applied to the policies, processes and procedures used by the organisation. These include but are not limited to the financial systems and reward and recognition frameworks used. In Apple, individual and team reputations act as a control mechanism.

Is your organisation tightly controlled? Which processes have the strictest controls? Are there incentives that promote collaborative practices? Do the controls inhibit collaboration and if they do—are they all absolutely necessary? Are projects structured and resourced to allow time and money is available for experimentation and risk-taking? Who are the people that receive promotions and bonuses, and which behaviours are rewarded by these processes?

4.3 Using the cultural web as an analytical tool

Some desk-based analysis of case studies is useful and indeed, it's the kind of study we will use here to illustrate the potential of the model. It can, therefore, be a useful tool for reflection about your own experiences. The Waters Centre for Systems Thinking, a not-for-profit foundation in the United States, has produced a useful set of courses and 'flip-cards' [17] which you might also find useful to prompt your reflections.

We want to make it clear, though, that for an organisational-level reflection with the potential to impact on colleagues, it is crucial to use a participative, or collaborative, approach. Any analysis of an organisation's culture also needs to focus on the *interactions between* teams and individuals and not on the individuals or teams themselves. If you draw in people from different parts of your organisation you are much more likely to be able to uncover the root causes of recurring patterns and, because of that, to come up with better solutions. There are many ways a system can run and reflecting on interactions rather than outputs can provide some space and leeway to make changes and progress.

Inevitably, if there is a recurring issue, it will mean that people will need to feel OK about acknowledging that some of the established ways of working are contributing to the problem. Conversations about preferred ways of working and established processes need to be handled in a psychologically safe environment with a focus on the problem that needs to be solved together. This cannot be rushed and is part of a process of organisational change; another topic about which much has been written and which we cannot cover in any depth here. For leaders and managers who want to implement an organisation-wide review of how they might address systemic issues impacting on collaboration, we would recommend engaging with specialist facilitators to help manage the conversations and reach meaningful outcomes.

4.4 Leading a collaborative organisation

'Unfortunately, many senior leaders view collaboration as a skill that is best applied on selected projects, rather than an organization-wide cultural value that should be embedded in the company's fabric.' [18]

In chapter 3 we wrote about the importance of team leadership in collaboration. Just as departments and organisations are systems, made up of lots of teams, the leadership of the organisations or institutions is critically important. Here, we are thinking primarily of the team of people that makes up the highest level of management in an organisation, typically called the Senior Leadership or Executive Management Team, or an equivalent group of people at a departmental or faculty level. We have already talked about the importance of values and guiding behaviours when we talked about creating collaborative teams and it is just as important when we look at the organisational context.

The global technology company Apple recognises that collaboration is core to both innovation and to its culture. It embodies this by encouraging cross-functional debate —it knows that every one of its products and its emerging technologies will be created by cross-functional collaboration. The case study of Apple's leadership by Apple executive Joel Podolny and academic Morten Hansen [19] describes the importance of leadership for the company and the need for its leaders to have what are commonly known as 'strong opinions, weakly held' or, as the authors put it, 'to be both partisan and open minded'. In Apple, these behaviours are facilitated by a deep understanding of, and devotion to, the company's values and common purpose.

The way an organisation is led will often define the culture of collaboration at the system-level. This is because leaders weave their beliefs, values, and assumptions into the high-level decisions through which the aims of the organisation are realised. Leaders in the top team need to embody collaboration as one of their values, create conditions where it can thrive and take decisions that embeds it into 'the way things are done around here'.

Research has identified a number of behaviours as being important for organisational leaders to adopt to develop a collaborative culture [20] (figure 4.5).

Pay attention: as the saying goes, 'where attention goes, energy flows'. Is the senior team paying attention to collaborative practice within the organisation?

Take a long-term view: senior leaders have to take critical decisions in the moment to manage immediate issues and incidents. Does risk management take account of the impacts of the risk and/or its treatment or mitigation on long-term collaborative arrangements and relationships, which are part of the organisation's social capital?

Be a patient investor: much like investing in stocks and shares, investing in collaboration is likely to provide a much more reliable return if you are in it for the long term. It is much less likely than some other forms of activity to provide a return within one or two financial years.

Act as role models: embodying the collaborative values of the organisation.

Recognise and celebrate collaborative behaviours as well as individual behaviours: both are important for collaboration and neither can exist in isolation in a collaborative organisation.

Recruit and nurture collaborative talent: does the organisation reflect its collaborative values in its top team?

Figure 4.5. Summary of the leadership behaviours needed in a collaborative organisation. After Callahan *et al* [20].

Be accountable: to whom is the top team accountable for investing in, modelling and rewarding the collaborative behaviours expected from all colleagues in the organisation?

4.4.1 Measuring and incentivising collaboration

As we have already seen, collaboration is a long-term investment in relationships. While we all want to have great outcomes from collaborative programmes, we need to shift the dial on the system, i.e. the actions, processes, behaviours, routines, the organisational habits and culture. If we can fix the inputs the outcomes will, most likely, look after themselves. James Clear, author of the book *Atomic Habits* [21], says that 'people who focus only on results win one time, people who focus on systems win again and again and so the place you want to focus is on building better habits and better systems, not necessarily on achieving a specific outcome'.

This way of thinking can give us a different perspective on what the most useful 'key performance indicators', or KPIs, might be for collaboration. We tend to measure the results of collaboration, which are highly visible and not the process of collaboration, which is often unseen.

'Leading' KPIs will measure change in the system and processes, lagging KPIs will measure whether a target, outcome or goal is reached. Too often organisations use lagging metrics to evaluate the performance of long-term collaborative processes. This is, in part, because it is hard to measure intangibles in a simple, quantitative way. Research by Heidi Gardner at Harvard Law School [22] has shown that when professional services firms collaborate to address complex issues, they gain a competitive edge. However, for individual professionals the financial benefits can accrue slowly, and other benefits are hard to quantify. This makes it hard for individuals to decide whether their investment in collaboration will pay off.

Leaders in collaborative organisations need to push for metrics which are meaningful in the context of collaboration, and that incentivise the systemic culture and habits that will move the organisation over time in the desired direction. This may mean that more qualitative approaches are needed to keep track of how the system is working with respect to research collaboration.

You will notice we have not said that leaders must be accountable for collaborative outcomes. We have noted that collaboration is a long-term investment by an organisation, and so leading measures are needed to ensure accountability for collaborative behaviours, rather than the lagging measure of outputs and outcomes. All outcomes are proceeded by process or practice, so finding measures of improved practice and ensuring those with the power to affect change are accountable for them, will go some way to help an organisation embrace a more collaborative culture.

4.4.2 Teaching collaborative skills

Francesca Gino points out in her article for the *Harvard Business Review* [23], 'Ask any leader whether his or her organization values collaboration, and you'll get a resounding yes. Ask whether the firm's strategies to increase collaboration have been successful, and you'll probably receive a different answer.'.

Gino goes on to observe that 'One problem is that leaders think about collaboration too narrowly: as a value to cultivate but not a skill to teach. Businesses have tried increasing it through various methods, from open offices to naming it an official corporate goal. While many of these approaches yield progress—mainly by creating opportunities for collaboration or demonstrating institutional support for it—they all try to influence employees through superficial or heavy-handed means, and research has shown that none of them reliably delivers truly robust collaboration.'.

In their white paper, 'Creating a Collaborative Organisational Culture' [24], Kip Kelly and Alan Schaefer make the point while that some of the skills required for collaboration are taught as part of team building or management training courses, to create an organisation-wide culture of collaboration, all employees must possess the skills to be able to collaborate.

Some suggestions for collaborative skills that can be taught include, but are not limited to:
- how to embrace change;
- how to ask for input from others;
- how to share information with others;
- how to listen for understanding;
- how to use negotiation skills;
- how to recognise and reward others;
- how to improve self-awareness; and
- how to reach consensus.

Skills audits, or needs analyses can be used to identify capability gaps. Once you know where the gaps are, it's much easier to put effective and targeted interventions in place, rather than a one-size fits all corporate programme.

4.4.3 Does everyone have to be in a collaboration in a collaborative organisation?

It's important to realise that we are *not* advocating that *all* leaders or researchers are themselves participating in a collaboration to be able to support the collaborative values of the organisation. Acting in line with the organisation's collaborative values in practice might look, for example, like:
- helping to reduce some of the barriers to other colleagues being able to collaborate within or between organisations;
- having a clear vision and purpose—where collaboration is a key value and guiding behaviour and a behaviour that's rewarded and celebrated;
- creating a culture of constructive challenge and diversity where a range of perspectives are welcome and people at all levels can speak out and contribute ideas;
- resourcing opportunities for colleagues to learn collaborative skills;
- providing seedcorn funding, including for travel expenses, to nurture emerging collaborations;
- reviewing workload models to account for the additional time and opportunity costs associated with collaborative practice.

4.5 Properties of systems that inhibit collaboration

Where systems produce emergent properties that get in the way of the outcome we want, we call this a 'system failure'. Understanding how some parts of the cultural web combine to produce effects that we don't want is important in figuring out where there may be leverage points to mitigate or address the failure. James Clear, author of the book *Atomic Habits* [25], puts this another way. He says, 'you do not rise to the level of your goals (your desired outcome, your target, what you want to achieve), but you fall to the level of your systems'. He says that if is there is a gap between your goal and your system, your system will always win, or, as the saying goes, 'culture eats strategy for breakfast'.

One of the easiest ways to identify where we have systems failures, or where we are falling to the level of our systems, is to use an approach advocated by Donella Meadows [26]—to ask 'what if?' questions.

We can use the cultural web to think of some 'what–if' questions for a collaborative organisation. For example, some what–if questions might include, but are not limited to: What if …
- There is not a shared vision and purpose throughout the organisation?
- Leaders do not 'walk the walk' of collaborative values and practice?
- Individuals don't understand how they can benefit from a collaborative culture as well as how it benefits the organisation?
- The opportunity costs of collaboration are too high for people to engage?
- Organisational control systems suppress emergent collaborations?

There are all sorts of systems failures that you might see, or have experienced, that impact on collaboration. We'll work through some examples that we have seen happen. We'll relate them back to the aspects of the Cultural Web, so you can see how useful this tool is for learning from what hasn't worked as well as you might have hoped.

4.5.1 Incentives that encourage loyalty to sub-organisational goals, e.g. local performance targets *(Cultural Web: Control system, stories)*

Locally optimising a team performance might not be the best way to achieve the organisation's highest priority goals. Sometimes compromises in local targets may need to be made to optimise the performance of the whole organisation. One leverage point is organisational goals—which can be phrased and communicated in a way that explicitly highlights collaborative processes and outcomes. Local targets might usefully be aligned to organisational goals and organisational targets, so that expectations are well-understood by everyone.

4.5.2 Weak organisation-wide understanding of the organisational goals, e.g. due to breakdown in communication of the goals; perception that these aren't relevant to someone of 'my' status/in 'my' team *(Cultural web: stories)*

This is a common problem in large organisations. It manifests when a team is not aware of how its goal and its members' personal objectives impact on both the overall organisational goals and the ability of the organisation to collaborate internally and externally.

Communicating organisational goals clearly at all levels can help introduce more openness and transparency into the organisational culture, which supports both collaboration and research integrity. It also supports problem solving and idea generation from a more diverse range of colleagues, improving the quality of solutions.

4.5.3 Lack of accountability for collaborative practice (*Cultural web: control systems, organisational structure, power structure*)

Leaders and managers in the system need to be accountable for collaborative practice (not outcomes *per se*), to improve the performance of the organisation as a system. If no-one is accountable for collaborative practice then people will tend to prioritise activities and outcomes which they are accountable for, often within their own team. This can be summarised as 'what gets measured gets managed'.

4.5.4 Poor cooperation between organisational units, e.g. between faculties, departments, professional services (*Cultural web: rituals and routines, control systems, stories, symbols*)

Incentivising cooperative processes aligned to organisational vision and mission can help promote collaboration across functional units. Training, such as work-shadowing in other parts of the organisation and mentoring by people who have successfully collaborated, can help break down barriers and help people see things from another's perspective. Celebrating role models and stories of collaborative success is another great way of demonstrating that 'this is the way we do things around here'.

4.5.5 Inadequate or poorly delivered feedback (*Cultural web: rituals and routines, control systems, power structures*)

The first step towards addressing a problem is to know that it exists. Expert in organisational learning and system change, Peter Senge [5], talks about the importance of identifying and addressing the root causes of problems in a system. Feedback to, and from, teams and individuals is an important part of this process and needs to be delivered sensitively to engage them in coming up with a solution, rather than feeling defensive or blamed. Being trained in how to give and receive the gift of feedback can be transformational for individuals and for organisations.

4.5.6 Organisational 'design' (*Cultural web: organisational structure, control, power structure, symbols*)

Often the way an organisation is set up is historic and has evolved over time rather than being holistically designed with the purpose or outcome in mind. A typical way that organisations deal with the structural silos that this often generates is to introduce a matrix of new cross-functional teams or structures. Unless a systemic review has identified that the organisational design is the key leverage point—these structures will be exposed to the very same system failures as the organisational system that they sit within and are likely to experience similar problems.

4.6 Putting the cultural web into practice—real world approaches to creating a collaborative organisation

In this section we will use case studies to demonstrate some of the aspects of the cultural web that often impact on collaboration and why these might be leverage points for you to think about. This isn't intended to be a fully comprehensive list, and every organisation's system is different, so it is important that you think about your own system or systems when thinking about your leverage points for collaboration. Nevertheless, we hope that illustrating these examples will give you some ideas and inspiration for your own thinking; you may even recognise these effects in your own organisation.

4.6.1 The impact of creating enabling financial structures and control systems

One of the recurring themes that came up when we talked to people about collaborating within universities is just how much of an impact the financial and legal structures, attitudes to funding, and processes have on collaborating across faculties and departments.

Sebastien Ourselin, Head of School, School of Biomedical Engineering & Imaging Sciences, King's College London told us about how they have developed their financial structures to facilitate collaboration.

'When you start on a long-term research collaboration you have to make sure that financially everyone has an incentive. To make this work in the traditional university you have to understand the impact of financial incentives from day 1. The beauty of what we have developed as a school at St. Thomas' Hospital is to bring the clinicians, the chemists, the engineers, the computer scientists and the biologists all in the same school so there is only one financial spreadsheet. So, the, 'what's in it for me' for every budget holder is easy to solve because there is only one budget holder. That solves a HUGE amount of problems. When people feel quite comfortable then they can collaborate. It's mainly about changing the model.'

If we look at this case study through the lens of the cultural web, we can see that there are a several things going on at St. Thomas'. Firstly, and crucially, they have amended the financial control systems to remove some of the barriers to collaboration. Secondly, there is a clear vision and purpose, 'making a difference to people's lives and making them better', that everyone is working towards and engenders a culture of collaboration. Thirdly, as Head of School, Sebastien tells a strong story about St. Thomas' as a place where collaboration happens, where people can feel comfortable to collaborate, and where people from different backgrounds work side by side without the distractions and inconveniences of worrying about satisfying multiple budget holders. In turn, this makes it easier for researchers at the School of Biomedical Engineering & Imaging Sciences to work with each other and with industry—a subject that we will come back to later in this chapter.

Of course, most of us work in organisations where it's not possible to immediately restructure and change the financial control systems. Collaborations have to sit comfortably within the organisational control systems and to satisfy the organisational requisite to bring in the research income that funds the generation of new knowledge. It's naive to think that we can have truly collaborative cultures if we do not have the money flows sorted; or that we will be able to sustain collaboration across departments or faculties if budget holders do not understand where their income will come from.

There can be really complicated financial control systems within large organisations like universities. There might be finance teams within individual faculties and/or departments, as well as at a university level. Finance teams might not be budget holders and often have different processes and performance standards (also known as key performance indicators, or KPIs) when compared to, say, academic researchers. Professional services focused on running efficient university finance, operations and estates might not be aware of why making collaboration more effective is important for the university. They may have no sight of the impact that interacting with multiple finance control points has on a researcher trying to collaborate across faculties or with partners like industry or other universities. And if we look at it from the perspective of the Finance Department, researchers may not be aware of their constraints, such as what is needed to satisfy an auditor. Exploring the control systems in a participative way, involving people from both sides could help to identify potential levers for change to make the university run more efficiently and with benefits for collaboration and research income.

One researcher told us their perspective:

'If I was running the university and I wanted to look at how to change the structure of how it runs to support activities of this type, I would have a finance team whose job it was simply to process quick-turnaround short pump-prime projects. In our faculty we have a brilliant finance team but trying to move things across faculties is really, really tough.'

'Why not have a cross-faculty finance team to fund interdisciplinary pump-prime projects? Instead of having a faculty team you could have a central rapid response team who would be able to mobilise if a funder needs a rapid turnaround from universities or for when you've got a pump-priming fund where money is handed out across the university. You would have a central team whose job it was to be fast response (and across the university) and that would take away the barriers.'

In a collaborative organisational culture you could have a participative discussion, bringing together researchers, finance teams and budget holders to explore options for change and to understand different perspectives. There are likely to be levers within the control systems and organisational structures which would be identified in a discussion. There may also be less obvious levers within the stories, rituals and routines of the teams and organisation as a whole—which could have a disproportionately positive or negative impact due to the emergent properties of the system. Some time and effort spent exploring these issues before making changes can really help to make sure that any interventions have the best chance of being successful with as few detrimental and unintended consequences as possible.

4.6.2 Creating the drive for multidisciplinary research and collaboration by changing the rituals and providing incentives

Simeon Yates, Associate Pro-Vice Chancellor for Research Environment and Postgraduate Research at the University of Liverpool told us about his experiences of developing interdisciplinary collaborations.

'I've seen loads of initiatives to develop interdisciplinarity. They run, say, a climate afternoon, and all the people interested in climate turn up. They all present on what they do on climate, and then they all walk away again. A lot of what comes of it is that they have a really exciting and interesting conversation, and some

people will find it interesting, some people will find it challenging, and some will want to take it forwards but there's no mechanism beyond lots of effort on their behalf. So, they go back to the lab or they go back to the office, and there's that paper to write. Then other people will say, 'Oh, we've invested in the interdisciplinary stuff, and nobody did anything.' Well, why would they?'

Here you can see how the rituals and routines around a collaborative workshop combine with the control system—the performance incentives—to get in the way of collaboration. People expect to turn up, have some interesting conversations and maybe spark some new ideas. But when they leave that event and go back to business as usual they quickly get sucked back into their usual routines through the control systems.

This is something that we have seen a lot when facilitating interdisciplinary workshops at universities. People come to workshops and both momentum and excitement are created but nothing happens. Liz will often talk to clients about the importance of thinking through how they intend to take things forward after a workshop and building on the momentum created, rather than letting the ideas, enthusiasm and motivation die. Some seedcorn funding can be transformational for legitimising and offsetting the costs of researcher engagement. So often organisations prioritise activities which are monitored and incentivised—winning grants, writing papers. The resources that are need for collaboration (time, money) can be hard to come by. The impact on the organisational ecosystem is reflected in the behaviours and activities of individuals and teams.

4.6.3 Symbols

In chapter 2 we talked about how important communication, particularly language, is to collaboration. Language is also part of the cultural web, being symbolic of your organisation, discipline or place. Simeon told us about how language differences impacted on his experiences running an EU project.

'We suddenly realised sort of six, eight months in, that we were using very different language when we got some of the first reports in, especially from some of the European colleagues whose English wasn't great. At first I just thought they're using the wrong words or they're using the words wrong because they were non-native English speakers. But suddenly I realised that we had thought we were having one conversation, but they've been having a completely different *conversation. When I understood what they meant by the words they had used, everything they had written made sense, but it wasn't at all what we had meant from the social science side!'*

'I think that actually doing that translation is how people get to understand each other's thinking. That takes time and that's why you've got to support people to get through that conversation in some manner or another. An institutional funding stream that supports that and gives people the time to do it is really important here in the University of Liverpool.'

Here Simeon has identified that the need to be able to translate language (symbol), needs an incentive (time, money). It's also clear that collaboration in the University of Liverpool is important to people who have budgets (power) and that it is an important part of their narrative (story) about the way things are done there. He's identified how to lay some essential foundations to shift the cultural paradigm towards one of a collaborative organisation.

4.6.4 Organisational structures

The London Nanotechnology Centre (LCN) is a UK-based multidisciplinary institute which is a joint venture between University College London, Imperial College London and Kings College London. Its purpose is to solve global problems in information processing, healthcare, energy and the environment through the application of nanoscience and nanotechnology. It's home to the laboratories of i-sense, the EPSRC-funded Interdisciplinary Research Collaboration in Sensing Systems. We met some of the researchers from i-sense earlier in the book.

LCN is a cross-organisational multidisciplinary unit set up to bring critical mass to the application of nanotechnology across many sectors, and to facilitate collaboration. It has a unique operating model that accesses and focusses the combined skills of all three universities across several key departments; Chemistry, Physics, Materials, Medicine, Electrical and Electronic Engineering, Mechanical Engineering, Chemical Engineering, Biochemical and Biomedical Engineering, and Earth Sciences. It also has strong relationships with the broader nanotechnology and commercial communities and is involved in many major collaborations, nationally and internationally. It's evolved in recent years to include research and innovation in quantum technologies.

Eleanor Gray, Mike Thomas and Isabel Bennett told us about their experiences of working within i-sense and LCN. As researchers working on a large multi-disciplinary project involving two of LCN's university partners, they have had a great deal of experience in navigating organisational structures which both encourage and inhibit collaboration. Eleanor, a biologist, reflected that at LCN, 'people from different cultures, disciplines and nationalities are all there, working together, because of their love of science and that's what drives them'. Isabel, another biologist, also told us how she hadn't realised how unique and enabling the set-up at i-sense was until she visited a prestigious US university and found herself as a biologist in an engineering group simply unable to do any microbiology *at all*.

All three postdoctoral researchers told us about how their experiences in i-sense and around the world had made them 'way more aware of making sure that everything is in place so that you can have access to the resources that you need for multidisciplinary research'. They had observed first-hand the pros and cons of the ways that different institutions, and different disciplines were set up and the logistics and practicalities of doing multidisciplinary research. Isabel summed it up when she said, 'I had taken it for granted in i-sense—but this is not the way lots of groups and universities are set up.'.

Lancaster University is a leading research-intensive university in the North of England. It has created its own interdisciplinary Research Institutes, to enhance interdisciplinary activities and to pull together critical mass in areas of excellence. They were set up to provide a formal framework through which its academic community could come together around key challenges.

If we look at Lancaster University's research institutes through the lens of the Cultural Web, we can start to get a picture of the leverage points that this university was using to change its collaborative culture.

Organisational structure	Lancaster introduced a formal change to its organisational structure to better configure it to support interdisciplinary research
Stories	The University has a strong narrative about its role as a leading research-intensive university. Its strong track record of high quality interdisciplinary research is a comparative advantage to protect and build upon.
	It tells a story internally and externally about its 'thriving ecosystem of interdisciplinary research' and collaborative approach, 'fostered by our mixture of formal and informal structures—including Institutes and University Research Centres—bringing together experts from different disciplines to address regional, national and global challenges'. The institutes were established to, 'build on Lancaster's strengths and culture', and to 'curate the interfaces where the best ideas can occur'.
Rituals and routines	Institutes have established their own sets of support structures to foster interdisciplinary activity, e.g. regular events and workshops, team building, a mentoring programme, research retreats and newsletters.
Symbols	Institutes created meeting spaces which were not perceived to belong to one particular discipline help to create a shared interdisciplinary space. Workshops and team building have been used to help develop a common language.
Power structures	The Research Institutes work with existing governance structures for faculties and departments to maximise overall benefit to the University, and for the delivery of collaborative programmes of research and impact.
	Institute Directors have an inclusive style of leadership and commitment to team building, a strong strategic vision and operational plan for the institute.
Control systems	Institutes are discrete accounting entities and are subject to the University's planning and approval processes.
	The Research Institutes Committee oversees the operation of the institutes and reports to the University's Research Committee. KPIs vary between institutes.

Lancaster University is using many of the aspects of the cultural web very successfully to embed and nurture the collaborative culture of the whole university, while pushing this model even further in selected multidisciplinary research areas.

These types of cross-faculty, single or multi-university institutes are becoming an increasingly common feature of the research landscape in the UK and globally. It is one way of having an organisational structure that is able to encompass both the needs of teaching and 21st century research, and we can see from the examples here that these can be very enabling for collaboration.

We would sound one note of caution though—that these types of structures are not, on their own, a magic bullet for enabling collaboration. All parts of the cultural web for the system these institutes sit in need to be thought about. These institutes and collaborations are exposed to the very same system as the organisations that they sit within or across and will be subject to the same pressures and system failures. The cultural web can be a useful tool for continually reflecting on the system(s) around and within the institutes, to help them fulfil their potential for enabling collaboration.

In our conversations with researchers, some highlighted how the use of publication KPIs (part of the control systems) can create some dissonance with a culture of collaboration. For example, when publication data collected and reported was attributed to the first author, some researchers we spoke to had experienced a tension between the team culture that had been created, and the individual need to have the publication track record in place to meet KPIs, obtain an academic fellowship or permanent post. One of them told us, 'it's not what anyone is about, personally, it's imposed by the system'.

These systemic tensions, which even multidisciplinary institutes are subject to, impact on relationships, which as we have seen are everything for collaboration. The Academy of Medical Science's work on team science [27] also highlights the need to address the whole system to support multidisciplinary research. We need to find ways of working, monitoring and measuring that support collaboration and don't erode the social capital that has been fostered. The cultural web is a great tool for digging down into these factors and identifying leverage points for change.

4.7 Thinking about collaboration, the cultural web and collaborating with industry

A collaboration between two different organisations is going to result in a collision of two systems. What happens when these two worlds collide will determine whether or not collaboration can happen.

An academic looking to work with a company is going to have to manage the intersection of two cultural webs—the web of the university and the web of the company. Just as a university has a heterogeneous mix of disciplines and cultures, companies are not homogeneous entities, and different industry sectors have different ways of doing things. Small and medium sized companies may have very different drivers to large, publicly listed multinationals.

Companies, and some other stakeholders, will often use or understand language that relates to the 'business, or value proposition', and 'the bottom line'. A business proposition is not some exciting research or papers in high impact journals. The power structures and influencers in businesses will primarily be concerned with how they can solve problems which will then go on to generate products or services that they can sell, or improving their processes to reduce costs or increase output, thereby increasing their profit margin. They need to stay competitive, act quickly and be first to market.

Imagine, then, that you are an academic who wants to approach a large company with a proposal for a long-term research collaboration. Where do you start? Liz has facilitated many industry–academic workshops and has some interesting learning which is relevant here.

As an academic, you might be excited because you can see the potential for this collaboration to generate lots of exciting new knowledge and high impact papers but know that, realistically, any new product or service is 5–10 years away. How does industry view your proposal? You may see it as transformational, but the company may struggle to see the problem you are trying to solve for them and how they might get a return on their investment. They might see this as a high-risk, low-return opportunity and your timescales could be too long. If that's the case, the chances of your idea being accepted and sponsored are slim.

From our experience, this is an all too familiar story. Liz has been approached many times by universities who are keen to show industry everything they can do. This might be via a symposium approach or showing companies the wonderful 'kit' that they have. Companies are interested in some of these pieces of kit and new ideas, but they are *much* more interested in how universities can work with them to solve problems. In our experience, the best approach to establishing a true collaboration with a business is to listen to them tell you what they would like solved with your help. This approach works much more often and can lead to a longer, more enduring and more transformative collaboration in the long term.

A great starting point for any industry–academic collaboration is thinking about your own perspective and that of your potential collaborator, to start to hone the 'business' or 'value' proposition for the collaboration. The value proposition is essentially a type of story which sets out the benefits of the collaboration and why it is worth investing in. You might like to look at the website of the company to see what the kinds of stories they tell about themselves are, and how your collaboration fits with their cultural web. Identifying how you can add value in this way will show that you understand the business and will differentiate you from other potential collaborations or collaborators.

Sebastien Ourselin told us about how he does his own homework for starting a new collaboration with clinicians and companies in the medical devices sector.

'The very last thing you should ask a company for when you start working with them is for any money. You don't go to them and say 'I'd like to work with you on a research programme in … and we start together from nothing. You need to be far more business-minded. You really need to understand where the clinical need is going to be and how this aligns with where the company is going. The company wants you to go to them understanding the need and the fit to their business, with some proof of

concept that is ready for taking to the next stage, for upscaling. They want you to have reduced the risk of the innovation for them.'

'So first, I will go and look at what the business/industry is doing today, and what they sell as the dream for the next 5 years. You can find this information quite easily online. Then, I will go to a few clinical journals to look at the trends in those areas.'

'I will use this background to make judgements about what the real limitations are in terms of engineering. Then I will go and see my clinical colleagues and start with what I have learned, rather than asking them what the next feature should be. I discuss with them the clinical needs and my ideas for technological innovation. I don't ask them what the next feature should be because they are focused on their responsibilities for looking after patients and delivering interventions. But that's the kind of clinician I want to work with because they are working with patients and want to do things that are safe, and we can bring them along with us on the journey. Of course, I try and influence them that I have a great idea—because I have done my homework—and I give them the opportunity to be part of it. When the time is right, they will help us test our technology and then we will have something which is ready for collaboration with industry.'

Here, Sebastien has two collaborations, each with a different purpose. He has a collaboration with clinicians, and its job is to understand the unmet clinical need, to shape the technological development, ensure that the technology is safe and to test the technology. His collaboration with industry has a different job, which is to address the issues of scale-up, manufacturing, regulatory approval and procurement so that the innovation can go from bench to bedside to start making a difference to patient outcomes in the real world.

Another control structure which is often cited as a barrier to collaboration by academics and businesses is that of attitudes to intellectual property (IP)—that is patents and licensing. Attitudes to, and processes for, prioritising and protecting protection vary at an organisational and individual level. Many researchers are frustrated by the time and effort that it takes to sort the IP requirements out. We have been told that in some cases collaborations have literally ground to a halt on a desk elsewhere in an organisation or that arrangements have taken so long to put in place that the research was no longer as timely as it had been. Everyone we spoke to acknowledged the importance of IP and its role in supporting innovation, but also noted that it had enormous potential to get in the way.

4.8 Some thoughts about the research and innovation system of systems in the UK

At the start of this chapter we defined a collaborative organisation as an organisation where there is a culture of collaboration both internally, across teams, and externally with people from other organisations. Each organisation is not only its own system, it sits within other systems. As our model showed (figure 4.2), an academic department sits within a university or research centre, which sits inside of a local innovation ecosystem, a national research and innovation funding system, and a global research and innovation system. We must not forget that organisations also sit within their places, their communities and broader society. They will be subject to influences from, and in turn influence, each of these systems and the way they interact with each other. Any action that we take to create a collaborative organisation will, therefore, be exposed to the same features as the system that they sit within and are likely to experience some of the positive and negative features of each of those systems.

The UK Research and Development (R&D) Roadmap [28] published in 2020 defines the 'R&D system' as:

'what connects universities, research institutes, government labs, charities and businesses to each other and to sources of funding. It supports the UK's wider ecosystem of public, private and third sector organisations to push the boundaries of knowledge and turn great ideas into economic, environmental and social benefits.'

The research and innovation system in the UK is really a system of systems. In the taxonomy of types of systems of systems [29], we can think of the UK 'R&D system' as a *collaborative system of systems* where 'the component systems interact more or less voluntarily to fulfil agreed upon central purposes. The central players collectively decide how to provide or deny service, thereby providing some means of enforcing and maintaining standards'. The central players in our system include, but are not limited to, the UK Government, research funders, universities, large companies, and publishers.

Just as we want to optimise an organisational system around agreed goals, the UK R&D roadmap sets out a desire from the UK Government to optimise some the features of our R&D system, highlighting in particular the need for this system of systems to be an 'efficient and effective system that enables strategic decision-making, provides the right incentives to enable research and innovation to inform each other, and ensures that money flows to the best researchers, innovators and entrepreneurs with the least friction possible', and that 'we should look to ensure that [the system] is coherent and efficient, with the right incentives for institutions to collaborate and not duplicate'.

The roadmap, therefore, sets out a clear aspiration for institutions to be able to collaborate within an efficient and effective system. However, the need to compete for scarce resources, and the need to convince funders that an organisation or research consortium can produce more value, over and above that of others, makes collaboration difficult in practice [30]. People that are in collaboration can also find themselves competing with each other for funds in order to satisfy different drivers or to address different aspects of a problem. Siv Vangen, Professor of Collaborative Leadership at the Open University in the UK, highlights that this is a paradox inherent in collaboration, that there are no single 'optimal' solutions to collaboration in practice, and that we need to recognise, accept, and manage the strengths and weaknesses of the different possible solutions. Realising the added value of a collaboration means that the interests, expertise, and resources of partners have to be brought together carefully within a very complex and dynamic environment.

The unwanted consequences of not recognising, naming, and managing the paradoxes of collaboration are especially found in large research and training grant competitions. Stakes are high, both in terms of money and reputation, for individuals and organisations. Collaboration is often mandatory and there may be the potential for teams to take a range of diverse approaches.

In such circumstances, tensions can arise within an existing collaboration where members find themselves competing with each other for new funding, perhaps as part of different consortia with dissimilar aims or areas of focus. The relationships that underpin collaborations have to hold space for these paradoxes and tensions and be supported by trust, transparency, and communication between partners. Funders and peer review have a role to play in creating a system that is more sympathetic to collaboration. For example, we have seen 'collateral damage' arising from a single competition for funding which have negatively impacted on trust and relationships which have taken years to build.

Within the UK research and innovation system, which as we have already seen is a system of systems, there are high levels of connectivity and the potential for emergent behaviour. This means that each participating organisation can look for leverage points which might shift the dynamics in ways which it considers desirable. Some of the interventions that result have been more effective than others. A great example of this is in the introduction of 'impact' into the assessment of research funding in the UK.

In around 2009, the then UK Research Councils (RCUK) introduced a new assessment criterion for the review of research proposals, to encourage researchers to be actively involved in thinking about how they would achieve 'excellence with impact'. Impact was broadly defined and included both academic impacts and socio-economic impacts. While researchers complied with the new requirements in terms of their proposal, this change in policy did not result in a significant cultural change within universities. What did lead to universities taking the so-called 'impact agenda' seriously, was the inclusion of impact as part of the UK Research Excellence Framework (REF) in 2014.

The UK REF is used to allocate nearly £2 billion in annual research funding. It uses peer assessments and research outputs across the UK as metrics for measuring success. Around 20% of a UK university's institutions income generated by REF 2014 would now be down to the quality of its impact case studies, and in total around £1.6 billion of public funds would be awarded on the basis of impact in the UK over the next five years [31]. This leverage point was so much more powerful than the one operating at a single grant level, and not just because of the sums of money involved, although that certainly was one factor. This leverage point worked at multiple levels in the cultural web. It engaged with the people with real power to make change within an organisation. The stories of impact, also known as impact case studies, became a way for the universities to communicate the value of their research to a broader audience outside of academia, and to different parts of their own organisation. Universities introduced control systems to identify and select the case studies that would be submitted as part of REF, and having a case study selected would be a source of prestige and therefore a symbol of your status and success.

The impact of REF on the UK R&D system of systems was highlighted in a speech by the former UK Science Minister, Amanda Solloway MP, in October 2020 [32]. She highlighted that, 'There are now very few parts of academic life in the UK that are not affected in some way by the REF', and that 'we must be prepared to look to the future and ask ourselves how the REF can be evolved for the better, so that universities and funders work together to help build the research culture we all aspire to'.

The challenge for all participants in our UK R&D system of systems is that it is complex and not all of its properties will be predictable and in line with the expected outcomes.

Which takes us back nicely to the first part of this book, the chapter on preparing yourself to collaborate. If you want to see an environment around you which is more collaborative, by preparing yourself to collaborate you can start to help make these changes happen.

The saying 'be the change you want to see in the world'[1] is apt here. The good news is that as part of the system, how we all behave matters. Small changes can have big impacts, particularly when carried out by lots of people, consistently, over time. If we want to create a more collaborative system, what we do as individuals, and teams, and organisations, matters.

---

[1] This is often attributed to Mahatma Ghandi. In fact, Ghandi said 'We but mirror the world. All the tendencies present in the outer world are to be found in the world of our body. If we could change ourselves, the tendencies in the world would also change. As a man changes his own nature, so does the attitude of the world change towards him. This is the divine mystery supreme. A wonderful thing it is and the source of our happiness. We need not wait to see what others do.' 'Be the change you want to see in the world' is a nice precis, or paraphrasing, of the quote.

4.9 Key learning points—chapter 4

- An organisation is made up of people, processes, behaviours, structures, skills and technologies, which combine to form a complex system.

- Complex systems have unpredictable emergent properties.

- Organisations sit within broader systems of multiple organisations.

- The Cultural Web is a useful analytical tool for identifying leverage points where an intervention might lead to meaningful and desirable change outcomes.

- The Cultural Web looks at the cultural paradigm through the lenses of stories, rituals and routines, symbols, power structures, organisational structure and control systems.

- For an organisation to have a collaborative culture, all employees need to have the skills to be able to collaborate. Not all employees need to always be part of a collaborative project; there are other ways to support and enable collaboration.

- Organisational leaders wishing to create a collaborative culture need to embody collaboration as a value, create conditions where collaborations can thrive and embed the value across the cultural web.

- Leading metrics will measure the process of collaboration, lagging metrics will measure the outcomes of collaboration. Because the impacts of collaboration are usually long-term, particularly for new collaborations, finding ways to measure and incentivise the collaborative process will support a more collaborative culture.

- Systems failures can be analysed and anticipated using what–if questions and the Cultural Web model.

- Collaboration between two or more different organisations leads to the intersection of at least two cultural webs.

- A starting point for a collaboration with a different type of organisation is to think about your own perspective and that of your potential collaborator, to develop a value proposition for the collaboration.

- The UK research and innovation system is a collaborative system of systems.

- Successful interventions often work at multiple levels of the cultural web.

References

[1] https://wellcome.ac.uk/reports/what-researchers-think-about-research-culture
[2] Handy C B 1999 *Understanding Organizations* 4th edn (London: Penguin)
[3] Meadows D H 2008 *Thinking in Systems: A Primer* (Hartford, CT: Chelsea Green)
[4] Vangen S 2017 Developing practice-oriented theory on collaboration: a paradox lens *Public Admin. Rev.* **77** 263–72
[5] See, for example Senge P M (ed.) 1990 *The Fifth Discipline: the Art and Practice of the Learning Organization* (New York: Doubleday/Currency)
[6] Duhigg C 2014 *The Power of Habit: Why We Do What We Do in Life and Business* (New York: Random House Trade)
[7] http://execdev.kenan-flagler.unc.edu/hubfs/White%20Papers/unc-white-paper-creating-a-collaborative-organizational-culture.pdf
[8] Johnson G, Whittington R, Angwin D, Regner P and Scholes K 2013 *Exploring Strategy.* 10th edn (Harlow: Pearson)
[9] Smith D, Schlaepfer P and Major K *et al* 2017 Cooperation and the evolution of hunter–gatherer storytelling *Nat. Commun.* **8** 1853
[10] HBR IdeaCast 2021 CEO Series: Mary Barra of General Motors on Committing to an Eco-Friendly Future https://hbr.org/podcast/2021/05/ceo-series-mary-barra-of-general-motors-on-committing-to-an-eco-friendly-future
[11] See, for example https://mckinsey.com/featured-insights/innovation-and-growth/telling-a-good-innovation-story https://forbes.com/sites/tendayiviki/2020/07/12/how-storytelling-can-help-build-an-innovation-culture/?sh=44e6130b62e8 https://hbr.org/2014/03/the-irresistible-power-of-storytelling-as-a-strategic-business-tool
[12] https://essentiallysports.com/rafael-nadals-rituals-the-mechanism-behind-the-rhythm-atp-tennis-news/
[13] Brooks A W, Schroeder J, Risen J, Gino F, Galinsky A D, Norton M I and Schweitzer M 2016 Don't stop believing: Rituals improve performance by decreasing anxiety *Org. Behav. Hum. Dec. Process.* **137** 71–85
[14] https://beautiful.ai/blog/beautifulai-presents-rituals-in-the-workplace, https://youtube.com/watch?v=4S84CzVQRy4&t=2s
[15] https://graphene.manchester.ac.uk/geic/about/
[16] https://ucl.ac.uk/healthcare-engineering/covid-19/ucl-ventura-breathing-aids-covid19-patients/our-team
[17] Thisisjude.uk and RAEng https://raeng.org.uk/grants-prizes/prizes/prizes-and-medals/awards/presidents-special-awards-pandemic-service/cpap-breathing-aids
[18] Podolny H 2020 How Apple Is Organized for Innovation. *Harv. Bus. Rev.* https://hbr.org/2020/11/how-apple-is-organized-for-innovation
[19] https://mckinsey.com/business-functions/organization/our-insights/tapping-the-power-of-hidden-influencers#
[20] https://thinkingtoolsstudio.org/cards
[21] Kelly K and Scheafer A 2014 Creating a collaborative organisational culture (IEDP UNC Kenan-Flagler Business School)
[22] Callahan S, Schenk M and White N Building a Collaborative Workplace http://anecdote.com/pdfs/papers/AnecdoteCollaborativeWorkplace_v1s.pdf
[23] Clear J 2018 *Atomic Habits* (London: Random House Business)
[24] Gardner H 2015 When senior managers won't collaborate *Harv. Bus. Rev.* 75–82

[25] Gino F 2019 Cracking the code of sustained collaboration *Harv. Bus. Rev.* 73–81
[26] http:\\execdev.kenan-flagler.unc.edu\hubfs\White%20Papers\unc-white-paper-creating-a-collaborative-organizational-culture.pdf
[27] Meadows D H 2008 *Thinking in Systems—a Primer* (White River Junction, VT: Chelsea Green Publishing)
[28] https:\\acmedsci.ac.uk\policy\policy-projects\team-science
[29] https:\\gov.uk\government\publications\uk-research-and-development-roadmap\uk-research-and-development-roadmap#being-at-the-forefront-of-global-collaboration
[30] Henshaw M, Dahmann J and Lawson B 'Systems of Systems (SoS)' in SEBoK Editorial Board. 2020 *The Guide to the Systems Engineering Body of Knowledge (SEBoK)*, v. 2.2 R.J. Cloutier (Editor in Chief). Hoboken, NJ: The Trustees of the Stevens Institute of Technology. Accessed [27-10-2020]. www.sebokwiki.org. BKCASE is managed and maintained by the Stevens Institute of Technology Systems Engineering Research Center, the Int. Council on Systems Engineering, and the Institute of Electrical and Electronics Engineers Computer Society https:\\www.sebokwiki.org\wiki\Systems_of_Systems_(SoS)
[31] https:\\kcl.ac.uk\policy-institute\assets\ref-impact.pdf
[32] https:\\gov.uk\government\speeches\science-minister-on-the-research-landscape

IOP Publishing

Research Collaboration
A step-by-step guide to success
Annette Bramley and Liz Ogilvie

Chapter 5

What's next for research collaboration?

'Work is not a place, it is what you accomplish together.'—Jim Kalbach, Mural

In this chapter we will explore the future of research collaboration and explore how you might integrate what has been learned during the pandemic into your collaboration practice. We'll look at how you can make the most of virtual or remote collaboration, hybrid environments and face-to-face interactions.

Finally, we will speculate about how other technological developments might support collaboration in the future and what this could mean for research leaders and skills for the coming years. As we write this, the pace of change and development is swift and what we write here may become out to date equally rapidly. Nonetheless, we have attempted to signpost you to what, at this moment, look like useful tools and resources, remembering that ultimately collaboration is about relationship, and relationship is about interactions between human beings.

5.1 Impact of Covid-19 and climate change on how we collaborate

When we embarked on the journey of writing this book in the autumn of 2019, a shift to more remote forms of collaboration seemed likely but still many years away. Fast forward 18 months, and the impact of Covid-19 on the world of work, including collaboration, has been transformative. Virtual collaboration tools have become ubiquitous, together with the 'you're on mute!' catchphrase of Zoom meetings. We have all been learning how to collaborate in new ways at pace and we will continue to refine and improve our use of the new platforms over the coming months and years.

Open innovation has been key to the advances of the last 18 months and is a trend that is set to grow. Sharing the coronavirus genome sequences online vastly accelerated vaccine development. Large consortia of companies and academics came together across the globe, with the shared purpose of saving lives. Whether the individual teams were sequencing the virus genome looking for mutations, developing possible treatments and optimising patient pathways, innovating medical technology to support patient care or coming up with new vaccines; the global challenge brought people together across disciplines and borders.

Clinical trials for the vaccines were filled at exceptional speed by citizens wanting to contribute to the global effort. Regulatory bodies were able to innovate within their processes to accelerate approvals without compromising patient safety.

The Nobel Prize winner Jennifer Doudna described the impact of Covid-19 from her perspective in an article for the Economist in June 2020 [1]:

'[Scientists] are working across disciplines, co-operating to make discoveries and applying novel technologies in a way reminiscent of the second world war. For example, in March my colleagues and I established a large consortium of academic and corporate scientists to create a clinical testing lab and to fast-track research on the pandemic.'

'This kind of multi-institutional group usually takes months, if not years, to build. But clearly it can happen much faster, to everyone's benefit, if barriers such as intellectual property are removed and there is a sense of shared urgency to solve critical global issues.'

It's hard now to imagine a complete retreat back to the pre-Covid ways of working.

As Doudna herself says: After Covid-19, science will never be the same—and this will be for the better.

This presents challenges and opportunities for collaboration, as a survey of office workers in the US, by PwC in June 2020 found [2]. The number one reason employees said they went into the office is to collaborate with other team members (50%). Difficulty collaborating was also the number one reason people give for being unproductive as they sheltered in place (39%), second only to balancing work with home duties such as childcare (38%).

The reality is that employees will not be returning to the same workplaces they left behind. There could be fewer people, restricted collaboration spaces and, in some cases, rotating shifts—all of which will require teams to find new ways to connect and collaborate. More than anything else, this need for connections is likely to shape what the workplace is going to represent.

We still face the huge global challenge of climate change. Research collaboration will be an absolute necessity to make progress, and at the same time, we will all need to adapt our own behaviours and practices to reduce our carbon footprints. Academics are a hypermobile community that can quickly rack-up the air miles. The total travel associated with attendance at a large academic conference can release as much carbon dioxide as an entire city in a week [3] and a participant at one academic conference can cause the same amount of carbon dioxide emissions as an average human during the course of an entire year [4].

The comprehensive uptake of digital collaboration platforms, and rapid progress on innovation in fields as diverse as travel, construction and energy, provides hope that we can reduce the greenhouse gas emissions associated with collaboration.

5.2 Remote collaboration

Before Covid-19, we had talked about using virtual platforms for collaboration but there was a persistent belief that face-to-face workshops would *always* be preferable for developing collaborations and building relationships. As the pandemic unfolded in March 2020 all face-to-face meetings, workshops and conferences were cancelled. We had no choice BUT to take to the virtual world to enable and maintain our personal and professional relationships and collaborations.

Tools for online collaboration that were available at the start of the pandemic were adopted at a staggering rate, and in the subsequent months even more apps and web-based interaction tools have become available.

Eighteen months on, remote collaboration looks set to remain part of our collaborative practice and we have learned an awful lot about collaborating online. It's important to remember that remote collaboration during a global pandemic where we all decamped to our homes almost overnight *is not and will not be* the same experience as remote collaboration under more 'normal' conditions. We have an opportunity to reflect and decide which parts of our new practice we want to take into the future, and what needs improving and refining to become sustainable, in all senses of the word.

So, what *have* we learned about research collaboration over the last 18 months?

Firstly, we learned that collaborations can adapt to, and thrive in, an online environment, but this doesn't happen on its own. Whether a workshop or meeting, we need to pay attention to the needs of different participants. People have been attending virtual events *in* new, shared virtual spaces and *from* different physical spaces whether that takes the form of a home office or a coworking space or a spare bedroom or a shared kitchen and many other spaces besides. The experience that is shared is a different one to that you experience when you are all physically co-present together. This needs planning for. As we adjust to the new ways of working and collaborating, we have needed *much* more time for preparation, to think about and create the experience and process that will deliver useful outcomes for the participants. We'll come back to this later in this chapter.

Secondly, and just as important is remembering that remote collaboration is still about relationships between human beings. The adage 'right people, right subject plus right process equals results' remains as true in the virtual space as the physical one. Collaboration is about getting to know each other, building trust, developing a common understanding and language and that is no different online than offline.

5.2.1 Synchronous and asynchronous collaboration

Collaborating remotely has some of its own jargon, including the terms synchronous and asynchronous communication.

Synchronous communication is real-time communication. A telephone call or a zoom meeting are forms of synchronous communication. A face-to-face workshop would also fall into the category of synchronous communication.

Asynchronous communication does not take place in real-time, which is to say that when a message is sent there is not an immediate reply. It means that not everyone needs to be in the same space at the same time. There are many digital tools for asynchronous communication such as text messages, emails, voice messages, chat channels, a shared online document, and whiteboard software.

For asynchronous *collaboration*—all you really need is a shared (usually digital) space for sharing thoughts and ideas, where team members can post reactions and comment.

There are advantages to online asynchronous discussions and collaboration. They:
- are good for avoiding group-think;
- provide space to think and respond rather than reacting;
- are time-efficient;
- allow for balanced discussions and give everyone a voice;
- are ideal for collaborations spanning global time-zones;
- are effective ways of sharing resources—put all the information in one place, which could include notes, chat (discussions), links, images, slides, audio/video clips.

Some companies have published, open access, their own virtual first toolkit [5] and ebooks [6, 7]. These contain resources and advice for remote working, communicating, and collaborating, including some great pointers about which tools to use when.

5.3 Increasing inclusion and flattening hierarchies

One of the most satisfying outcomes of the move to more virtual collaboration platforms has been the impact on equality, diversity, and inclusion in research collaboration. The move to remote working has meant that a more diverse range of participants can attend workshops. For example, those with disabilities or caring responsibilities that might have struggled to travel to a face-to-face workshop can attend from home or another venue that is suitable for them. Virtual content can be recorded and shared to increase participation and to enable those that have had to miss parts of synchronous discussions to catch-up, and feed in their ideas and perspectives.

We have noticed an increase in the number of women, early career researchers, people from minority ethnic backgrounds, introverts, neurodiverse people, people with physical disabilities, people with lived experience and community representatives able both to attend workshops and relishing the opportunity to volunteer their opinions and thoughts. These people are now part of collaborations that previously they might not have had the chance to participate in. This has to be a good thing for research and innovation. Digital exclusion is an important issue, particularly in less affluent communities. In general researchers and their collaborators have access to digital collaboration tools and can facilitate the participation of others.

We have also noticed an increase in participants from countries all around the world at remote workshops and conferences. The world, literally, is at your fingertips. It is now easier to build and maintain a relationship with an international partner, or to have co-supervisors and mentors located in other countries. This has the potential to being new global perspectives, new routes to impact and access to different networks to researchers wherever they are based.

In early 2021, Liz facilitated a series of international sandpits for the British Academy, on the themes of 'Just Transitions', 'Global Disorder' and 'What is a Good City?'. Virtual spaces enabled people from India, Columbia, and Brazil to attend, sparking great conversations between participants from the global North and global South. While it would have been possible to have held this face-to-face, it all happened more easily because the event was virtual. Holding the workshops in the morning helped to accommodate the different time-zones for this particular group of attendees. Keeping the workshops shorter because of the virtual environment also worked well for those attending from outside the UK.

Wellcome's Reimagine Research festival in March 2021 was their most inclusive event ever [8], including British Sign Language (BSL) interpreters for those with hearing impairments and publishing a code of conduct online in advance. The National PostDoc conference 2021, hosted by the University of Liverpool employed BSL interpreters from their student body to support the event. There is no one-size-fits-all when it comes to accessibility, and relatively simple adjustments can make a huge difference. Even the dates you choose are important. For example, the various countries in the UK have different timings for the school year and public holidays and this can particularly affect those with caring responsibilities. It is worth thinking in particular about the dates of public holidays in different countries if you are facilitating international collaborations.

There is a lot of great advice about making events accessible available online [9], which we will not try and duplicate here if only because it is sure to date rapidly! Instead, we encourage you to do your own research and seek advice locally about the resources available to you.

The dynamics of online meetings seem to flatten hierarchies, and the influence of the 'highest paid person's opinion', or HIPPO, seems to have been reduced. For some, the online environment has provided an atmosphere that allows them to feel more psychologically safe and more able to express their ideas and opinions. This may be due to the way that we are displayed on screen, inside similar sized boxes, or to the dynamics of a well-facilitated online discussion. With that said, the person chairing or facilitating the meeting has an essential responsibility for sharing power and helping to ensure that everyone feels heard.

As most people have been participating from a home-base and have opted for more casual work wear, many of the 'pseudo-status markers' have disappeared. We see the merging of home and work with dogs barking, cats' tails appearing, children joining meetings and of course the doorbell heralding the arrival of the postman or an online order.

Helen Szoor-McElhinney of the University of Edinburgh told us about her experiences working remotely with community partners as part of the 'Our Health Interdisciplinary Community-University Research Programme' [10]. Feedback from the community partners included that they felt more confident to express their opinions because they felt safe within their own home. Their physical comfort contributed to their psychological safety, as well as being surrounded by familiar objects, sounds, and smells. Independent control of the volume level was also a huge benefit for accessibility.

We have also had the unique opportunity during various lockdowns to share more of ourselves with others. We often see that others are not so different from us after all. Brent Ziegler of Dyer Brown Architects in Boston noted in an article for Forbes.com early in the pandemic:

'While remote work reinforces separateness, it also offers a kind of intimacy now that we are all invited into each other's personal spaces and home offices.' [11]; and

'...there may be a new sense of connectedness not seen in more sterile office settings.'

Our collaborations will be much richer if we can create and maintain inclusive environments. After all, most of us are likely, at some point in our lives, to benefit from certain adjustments that can be made to make it easier and more enjoyable to attend events. Creating inclusive environments will help create psychologically safe spaces, so important for creativity and collaboration, and in turn create spaces where excellent research can thrive.

5.4 Planning remote workshops

In this section, we will cover some of the lessons that we have learned from planning and attending online workshops. There is an enormous amount of flexibility in how you approach an online workshop, but some over-arching principles around the structure remain. These are similar to a face-to-face workshop, with the remote or online environment bringing new opportunities without the dreary or mind-numbingly dull environment of conference meeting rooms!

We both learned very quickly that a whole day or half-day workshop online with a group of people is exhausting and does not lead to the best outcomes. We had better results when we broke the online discussions into shorter 'sessions', which lasted for a maximum of 2 h. A workshop then becomes a composite of synchronous 'sessions', interspersed with opportunities for asynchronous collaboration.

We can think of a typical workshop as taking place in a number of sequential phases, each with a distinct intention.

Phase 1: The start of the workshop, getting to know who is online, introducing the purpose of the gathering, the background and context. This would typically be a plenary session and could incorporate some online breakout rooms for introductions, much like you would have smaller group discussions around tables in a face-to-face event.

Phase 2: A crucial phase of exploring the challenge and the constraints together. This is a great point to bring in people with lived experience, and/or the workshop sponsor or problem owner. This is a time to ask open questions and listen to understand. An online workshop gives much more prospect of bringing people 'at the coal face' into the virtual room than a physical workshop. An advantage of remote workshops is that there is more opportunity to thoroughly explore and understand the nuances of the problem space and to pose better questions which, in turn, can lead to more innovative and applicable solutions.

Phase 3: The group works together and/or as sub-groups to come up with and to refine ideas. This phase is a mixture of synchronous and asynchronous activity, moving between whole group and sub-group discussions, and feedback to share, refine and build on the ideas that have been generated. A further advantage of online workshops is that you can allow plenty of thinking and interacting time between plenary sessions and extend a workshop over a longer duration than would be possible in a shared physical venue. As we have already seen, online workshops can also be more inclusive.

Phase 4: The workshop closes, with a clear review of what has been achieved and what the next steps will be.

In schematic, the timeline of a workshop then might look something like that shown in figure 5.1.

This isn't a rigid formula—but a framework which you can play with to suit the needs of the group you are working with. A group whose members already know each other well will not need as much time for introductions, and you will be able to focus synchronous conversations on specific areas which will ensure that ideas and solutions are innovative and useful.

You might choose to help attendees start thinking about the context for the discussions by sending a link to some reading material or a TED talk on YouTube with the joining instructions—this is pre-meeting asynchronous communication.

Just because we are collaborating remotely doesn't mean that all of our activities and creativity have to be expressed virtually. The camera on your phone and your webcam are handy tools for converting analogue to digital. You could use sticky notes to brainstorm and share them with a digital photo taken on your phone. Walls and windows, even a piece of cardboard can be transformed into a whiteboard, with the right resources and a bit of forethought. You can invite participants to take a walk for fresh air to reflect or for a 'walk and talk' by phone. We have captured

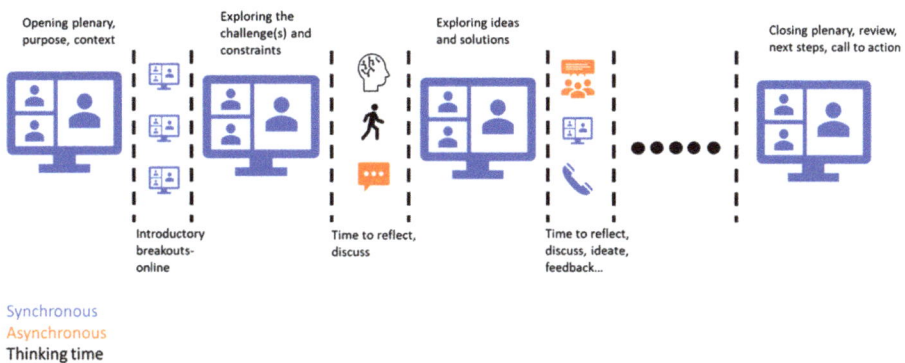

Figure 5.1. Schematic diagram to show how synchronous, asynchronous and thinking time segments could combine in a timeline of a virtual workshop.

voice notes while walking and brainstorming and then typed them up into powerpoint for a feedback session online. Mixing things up helps to keep people involved with an event, and can help increase the range of engagement cues that we experience.

Resources can be dispatched to participants ahead of time, to make sure that everyone has the tools that they need whatever the physical space they have access to. You can even provide teabags or biscuits for that real workshop feeling! Mixing things up can help to keep the workshop unique, memorable, and with a playful, light, and psychologically safe ambiance that can help new ideas to come through. Be sure to choose processes that are appropriate for the group you are working with, so that you can create an equal space where all your participants feel able to contribute.

5.4.1 Some tips for remote workshops and virtual events

Zoom less, zoom better [12] could be the mantra for our times and is true for remote workshops and meetings alike. Keep online participation to a maximum of 2 h at a time. In the second chapter of this book, we explained the importance of non-verbal communication, and that 'Zoom fatigue' arises from the absence of the non-verbal signals we are well adapted to pick up. In his e-book, *The Science of Virtual Engagement* [13], speaker and consultant Scott Gould goes further to say that the scarcity of engagement cues in online environments actually 'disrupts the process of neural synchrony' whereby we adopt similar brainwave patterns to the people we are face-to-face with.

Limit, presentations or broadcast communications to a maximum of 10 min 'talking at' the participants: during online presentations, we have all seen a screen full of engaged faces turn into a screen of initials, black boxes with names, or photographs. We have noticed that it appears as though our natural limit for listening to understand an online presentation is around 10 min or less. It's better to keep broadcast communications short, punchy and effective, interspersed with time for smaller groups to reflect on what they have heard.

It is very easy to enable conversations in breakout rooms and bring people back to plenary for a Q&A with the speaker, which is also likely to generate more interesting and relevant questions, as well as better embed the understanding among participants. Interaction software can be used to get feedback from larger groups, but be aware that there is a risk that people disengage and become distracted if they have to switch between devices.

Use breakout rooms to enable discussion and knowledge sharing among participants: online platforms have made breakout discussions so much easier and quicker. As the platforms have matured it has become possible to offer participants a choice of breakout rooms, to allocate people to specific breakout groups or to simply randomise participants. What format you choose will depend to a large extent on what phase of the discussion you are in, and the extent to which participants know each other. Use the chat function to remind breakout groups of the question, task, or purpose of the breakout: this can really help groups stay on task!

Avoid death by debrief: although online breakout groups are easy to form, when every group feeds back on the discussions they have just had you can just as quickly lose the momentum of the event. You can use asynchronous tools to capture outputs. A slide capturing a few bullet points can be shared online to feedback, and can be circulated by email or saved to a shared space in the cloud. Photos are another great way of capturing ideas generated on sticky notes or paper. Voice notes are gaining popularity as a way of communicating asynchronously and could be a great way to summarise discussions.

Breaks in the session can be used to provide a form of virtual marketplace when individuals can browse the outputs, and send through questions or observations to facilitators be picked up at the start of the next session.

Breakout room hosts are not essential but can be helpful: in a face-to-face workshop it would often be possible for facilitators to spot groups that are struggling with the question or going 'off task' or for groups to grab a passing facilitator for help. Some platforms have a 'help' function for breakout rooms, which is a great innovation. Depending on your group and the phase of the discussion, you might find it helpful to identify a participant or a facilitator to 'host' the breakout. Having a room host can also help ensure that all of the voices are heard, not just the dominant ones.

5.5 Engineering serendipity into online collaboration

One of the biggest challenges of online collaboration and workshops is how to create the random bumping into one another—the serendipitous lunch or coffee queue conversations that can lead to cross-fertilisation of ideas and expanded networks. For many years we have been talking about the importance of investing in social capital for multidisciplinary research and innovation. There is a sense that during the pandemic we have been expending the social capital that we had built up prior to the lockdowns, and at some point soon we will need to top-up our reserves.

There have been many articles and blog posts written during the pandemic that try to articulate the intangible benefits of being co-located with our co-workers and collaborators. There is a tangible sense of loss—of relationship, of camaraderie and of the 'buzz' of being around people that share a common purpose. Human beings are social animals after all. We will need to design our time together differently to maximise engagement, whether through virtual or face-to-face events and meetings.

In a blog about hybrid collaboration [14], Jim Kalbach of the online collaboration software company Mural reflects that:

'There's an important social component to how people come together in a physical setting (as the isolation of the pandemic made abundantly clear). Working together in the same space fosters connections between colleagues and builds a strong company culture. ... we surveyed over 400 people on their experiences with remote collaboration. Their biggest frustrations? Missing social connection with colleagues, lack of spontaneity, difficult communication, and struggles with creativity (among others).'

Earlier in the book we wrote about the importance of serendipity, and how you can prepare yourself to experience more serendipity by putting yourself out there, sending out a flare, being a go-giver, showing up fully and staying open. This is still true in an online event. You can prepare for serendipity in an online environment, or, if you are the event facilitator, prepare your participants for serendipity.

Many online meeting platforms now have spaces for deliberate mingling or informal conversations. The most obvious of these is the chat function—introducing yourself in the chat and engaging with others in the room can be an easy way of encouraging serendipity. Most meeting platforms will assign people to smaller groups or breakout rooms. Using the random function is a very quick way to mix people and can be used very effectively in a form of group 'speed dating' to create lots of interactions with different people.

There has also been a proliferation of online networking tools which provide a virtual environment for networking. One of these platforms has been described as [15], 'a mashup between Zoom and an 8-bit videogame. It allows you to have multiple separate video chats at the same time. You can enter and exit them, just as you might at a real party or event, except here you navigate a pixelated avatar around one of its virtual worlds.' Other platforms allow you to upload a document so are great for a marketplace feedback, or a facilitated activity with the added bonus of serendipitous encounters. An online 'water cooler' [16] is an asynchronous digital space where you can leave notes and comments and connect with your co-workers. It's essentially a social online whiteboard. You could also use or other social media platforms to create this kind of interactive experience.

Lynda Grattan, Professor of Management Practice at London Business School, says in an article for *Harvard Business Review* [17], that for innovation (or what we might call creative thinking or ideation), this will often be:

'*stimulated by face-to-face contact with colleagues, associates, and clients, who generate ideas in all sorts of ways: by brainstorming in small groups, bumping into one another in the hallways, striking up conversations between meetings, attending group sessions. This kind of cooperation is fostered most effectively in a shared location—an office or a creative hub where employees have the chance to get to know one another and socialize. To that end, cooperative tasks must be synchronous and conducted in a shared space. Looking to the future, we can expect that the development of more sophisticated cooperative technologies will render shared* physical *space less of an issue.*'

5.6 Future directions for research collaboration?

Looking to the future, it's likely that next generation collaboration tools will become much more sophisticated. Google's Project Starline [18] aims to create technology that might be able create the feeling of being together with someone, just like they're actually there—life-size and in three dimensions. You would be able to talk naturally, gesture and make eye contact with your collaborators. AltspaceVR is a social platform software, now owned by Microsoft, where individuals can gather, talk and be co-present in large groups. You need virtual reality (VR) hardware for the full experience, but there is also a 2D mode which enables people without VR headsets to participate.

Within a few short years, virtual collaboration could mean meeting and working together in photo-realistic immersive environments, as natural and intuitive as meeting face-to-face. These environments could also be used to help researchers experience a particular problem 'first-hand', exploring it from a range of angles and perspectives to really understand the challenge.

There are signs that artificial intelligence (AI) could be used increasingly to help customise team building. Even now, platforms are available to businesses that use AI to cluster employees in an office, by work style or collaboration profile, to create more opportunities for them to interact.

Fran Katsoudas, CISCO's Chief People Officer describes how AI is transforming team building and collaboration at CISCO [19]:

'AI and machine learning is helping us better understand how our people think and work. It's helped us develop perks to incentivize our employees, find pools of hidden talent around the globe and develop new ways to stimulate innovation.'

'We are learning new ways to collaborate and team build.'

One thing the company has found is that like-minded workers gravitate towards each other, aided by technology, sparking bubbles of innovation. AI has also helped CISCO 'make breakthroughs' in hiring and talent development.

'Now we are doing what we call 'blind hiring', where we don't see the name or the university the candidate attended. It's helped eliminate bias and lets us just review the purity of a person's work. This has opened up a whole new pool of talent we can tap—individuals who may not have a college degree but are skilled at coding and a host of other expertise.'

Data analytics will be actively used in some large organisations to form teams and identify the 'right' talent for projects, and potentially to find 'hidden' talents and skills. If you are thinking that this all sounds a bit dystopian, our view is that AI is unlikely to be able to perfectly match people for research collaborations any time soon.

It is one thing to use AI and machine learning to select the team members for a specific, well-defined project, based on a database of skills, experience, and qualifications. As we saw in the first chapter, there is a significant difference between teamwork and collaboration. Teamwork requires individuals to complete certain tasks that together deliver the team's end goal. Collaboration requires an alignment of values, co-creation of solutions and supportive interpersonal behaviours. We haven't figured out how to measure collaborative behaviours properly yet, and so the training data for collaboration just isn't available in the way that it might already be in large organisations, like CISCO, that are looking to pull functional teams together.

There are significant equality, diversity, and inclusion challenges involved in the use of AI and machine learning for identifying people for research collaborations. Structural inequalities which exist today will be reflected in current data, and we will need to be mindful of this in training algorithms. Human beings do have unconscious biases, so a positive use of AI could be to identify a broader base of people for workshop and conference invitations, or for 'brain dates'—those which would normally have not been 'on our radar' otherwise. This potential for AI to augment human beings to improve inclusion can only be realised IF the algorithms and training data themselves do not contain inherent biases.

It is one thing to speculate about the use of AI and machine learning to help derive a long list of workshop participants, where the interactions between human beings during a facilitated process and the relationships that result will support collaborative outcomes. This seems to us a potentially helpful use of this technology. When it comes to the practice of collaboration there are more complex factors at play. These include, for example, the number of variables that have to be taken into account, the difficult to quantify nature of collaborative behaviour and the deep social and linguistic connections that we make when we develop our new tribes, combined with the not-always-logical nature of interpersonal dynamics. While there may be a place for AI and machine learning in the process of collaboration-building, it is unlikely that these technologies will be able to successfully 'match-make' collaborative research teams any time soon.

Over the next five years, the transition to remote-first ways of working and collaborating will be supported by a plethora of new and more sophisticated hardware and software technologies. What will not change is that on the other side of the hardware and software is a human being. Getting into collaboration with another human being requires all the important skills and capabilities that we have outlined in the previous chapters. These are future-proof skills that will help you thrive and make the most of the new technological platforms that come online in months and years to come.

5.6.1 The world of hybrid collaboration

Covid-19 has had a huge impact on the way we all learn, work, socialise, and collaborate. Many of us have experienced the advantages of being able to have more control over *where* we learn, work and collaborate.

The last 18 months have seen a decade's worth of transformation of how we work and collaborate. Things will continue to evolve, and rapidly, as new and better solutions become available that enable remote and hybrid collaboration, both in offices but also in laboratories and other settings.

The 4th Industrial Revolution is going to lead to more automation; Professor Andrew Cooper's group in the University of Liverpool have already demonstrated a mobile robot that can run experiments 21.5 h per day (stopping only to recharge its battery) [20]. The robot automated the role of the researcher, screening thousands of samples and making fewer mistakes than a human researcher. In the future, these kinds of technologies will free human collaborators up to think creatively and imaginatively, and to 'play' in the lab rather than to grind through repetitive tasks. In principle, a researcher somewhere else in the world, and in a different time-zone, would be able to carry out experiments in a lab manned by robots.

This all points to increasing flexibility and diversity in how collaborations start and develop and, yes, come to an end. We will move from virtual to face-to-face to hybrid seamlessly, making the most of our time physically together. There will be a lot of learning about what works and what doesn't work along the way. As the new world emerges, organisations and employees are beginning to embrace the opportunity to rethink how we work, with many indicating that a hybrid model is the way of the future.

In an article for *Harvard Business Review* [17], Lynda Gratton of the London Business School outlines a model for optimising hybrid work focussed on the axes of place and time. She points out that place is the axis that's getting the most attention at the moment—will we return to the workplace or won't we? Less noticed is the shift many people have also made along the time axis, from being time-constrained (working synchronously with others) to being time-unconstrained (working asynchronously, whenever they choose).

Applying this model to research collaboration, particularly facilitated collaboration, also leads to some interesting insights.

Prior to Covid, workshops to initiate collaboration would happen on a defined day in a defined place. Now, collaboration can be initiated online (from anywhere) synchronously or asynchronously (any time), and continue from anywhere at any time, using digital tools.

What becomes important, in Grattan's model, is to think about how best to collaborate from four distinct perspectives:
- jobs and tasks,
- personal preferences,
- projects and workflows, and
- inclusion and fairness.

Within a research collaboration, people with different roles will have different needs to perform each of their tasks effectively. She also highlights the importance of stepping back and ensuring that whatever arrangements you put in place support the values and the culture you want to create.

One way to do this might be to refer back to the cultural web model in chapter 4. Have you created a hybrid way of working that supports the collaborative culture that you want to see in your team or your organisation? The Cap Gemini 'Future of Work' report [21] says that we will need to reinvent a trusted work culture with new collective rituals, with 68% of employees preferring their organisation to focus more on team building activities and 61% wanting their organisations to recreate informal connections innovatively, through virtual coffee chatrooms or virtual water cooler moments.

5.6.2 Leading hybrid collaborations

In chapter 3, we described the leadership behaviours that support research collaboration and the responsibility of leaders to create the conditions for collaboration to thrive. In the collaborative leader model, the key behaviours are connecting people, attracting diverse talent and modelling collaboration.

In a 2019 article, author and collaboration expert David Coleman highlights that the biggest problem for sustainable distributed collaboration is the behaviour of groups of people, also called 'culture' [22]. He says that without a collaborative culture and direct support and participation from 'management'—who we would think of as leaders—no matter what tools and processes you have, it will be hard to sustain distributed collaboration.

Whether in a virtual or a physical environment, leaders will need to continue to communicate the vision and role model the behaviours consistent with the values and culture of the team and organisation. More than ever, they will need to be consistent in how they do this in different environments to drive trust and engagement across the collaboration. Emotional intelligence was an important part of a research leaders' toolkit before the pandemic; in a hybrid world it will be essential.

Leaders will also need to embrace the collaborative tools, environments and processes adopted by the team, to create a culture that supports ongoing collaboration and to step in if they see that the quality of the relationships and interactions are dropping off.

5.6.3 Physical spaces for hybrid collaboration

As we move into a new world of hybrid working and collaboration, we will need to rethink how we organise our work environment, how we integrate virtual and physical spaces and participants in them and how people will interact in these environments. The same is true for research collaborations and workshops.

The title of an April 2021 article by academics Anne Laure Fayard, John Weeks and Mahwesh Kahn sums this up nicely: 'Designing the Hybrid Office— from workplace to culture space' [23]. The authors rightly highlight the important role of 'human moments' within workplaces, the importance of relationships not just acts of collaboration, and how frequent in-person interactions lead to commitment, support, and cooperation among people in teams. They say that physical cues and non-verbal communication will enable a richness of understanding that is not possible in a virtual space and will give us much more insight into what others are thinking and feeling, further strengthening relationships, and building trust. The warmth and energy of face-to-face conversation could help to increase motivation and team spirit.

Physical spaces that support structured and unstructured collaboration will be an important part of our hybrid working environments. They will be designed to support the kinds of interactions that cannot happen remotely, and to reinforce the values and cultures of our teams and organisations. They will, in effect, be social hubs, with informal shared spaces for spontaneous and relaxed conversations. They will provide a setting for experiences that bring people together, in some cases literally 'breaking bread' together. Non-work-related rituals are a great way to reinforce the culture and signal that getting together physically is about connecting as well as collaborating. We saw in the previous chapter that physical spaces, rituals, and routines are an important part of creating a collaborative culture; and will be even more important as we adjust to hybrid working.

If you are lucky enough to be in the position to design a physical space for hybrid research collaboration—what else could you usefully think about? A combination of more open and more private, or 'huddle' spaces will be handy. Perhaps think about privacy through the lenses of acoustic, visual, territorial, and informational 'leakages'. Spaces for bringing in those that can't be with you in-person will be essential, and it looks likely that technology for virtual and augmented reality will move more quickly into workplaces as a collaboration tool over the next few years. The accessibility of meeting places using public transport will become increasingly important as we look to decarbonise our collaborations.

Many of us have increased our appreciation of the natural environment during lockdown, and you might consider leveraging the outdoor spaces available to you as 'pop-up' breakout rooms or for 'walkshops'. Biophilic elements like plants and water offer sound, colour, and texture. These can inspire creativity and can be used to provide privacy and define boundaries inside and outside buildings as well as having a role in capturing carbon and supporting biodiversity.

5.7 Key learning points—chapter 5

- Remote collaboration is still about relationships between human beings. The adage 'right people, right subject plus right process equals results' remains as true in the virtual space as the physical one.

- Remote workshops need as much if not more preparation, to create the experience and process that will deliver useful outcomes for the participants.

- *Synchronous communication* is real-time communication; a*synchronous communication* is when a message is sent and there is not an immediate reply.

- Remote workshops can be inclusive and flatten hierarchies, as long as attention is paid to what participants need to feel psychologically safe online.

- The best outcomes were observed in online discussions which were divided into sessions each of which lasted for no more than 2 h.

- A workshop progresses through a series of phases which can be carried out asynchronously or synchronously.

- For great online workshops: zoom less, zoom better; keep broadcast communications short; use breakout rooms; avoid death by debrief; consider having breakout room 'hosts' and record only for the record.

- We have to be more proactive in 'engineering in' serendipity in online environments than when we meet face-to-face.

- More sophisticated collaboration tools and other technological developments are likely to transform remote and hybrid collaboration in the next decade.

- Research leaders will need to continue to communicate the vision and role model the behaviours consistent with the values and culture of the team and organisation.

- Physical collaboration spaces will be designed to support the kinds of interactions that cannot happen remotely, and to reinforce the values and cultures of our teams and organisations.

References

[1] https://economist.com/by-invitation/2020/06/05/jennifer-doudna-on-how-covid-19-is-spurring-science-to-accelerate

[2] https://pwc.com/us/en/library/covid-19/assets/pwc-return-to-work-survey.pdf

[3] Klower M, Hopkins D, Allen M and Higham J 2020 *Nature* **583** 356–59

[4] Jäckle S 2021 Reducing the carbon footprint of academic conferences by online participation: the case of the 2020 Virtual European Consortium for Political Research General Conference *PS: Pol. Sci. Pol.* **54** 456–61

[5] https://blog.dropbox.com/collections/virtual-first-toolkit

[6] https://mural.co/blog/6-keys-to-collaboration

[7] https://virtualnotdistant.com/

[8] https://linkedin.com/pulse/how-we-organised-our-most-accessible-event-ever-erika-loggin/

[9] Just one example is https://visitscotland.org/binaries/content/assets/dot-org/pdf/marketing-materials/accessible-events.pdf

[10] https://ed.ac.uk/clinical-sciences/our-health

[11] https://forbes.com/sites/joemckendrick/2020/04/19/serendipity-lost-how-all-digital-engagements-reshape-innovation

[12] Syed M May 2020 We can go on meeting like this *The Sunday Times*

[13] https://info.swapcard.com/virtual-event-engagement

[14] https://mural.co/blog/why-hybrid-collaboration-is-harder-than-you-think

[15] https://wired.com/story/zoom-not-cutting-it-virtual-world-online-town/

[16] https://mural.co/blog/virtual-office-water-cooler

[17] Gratton L 2021 How to do hybrid right: when designing flexible work arrangements, focus on individual human concerns, not just institutional ones *Harv. Bus. Rev.* **99.3** 66

[18] https://blog.google/technology/research/project-starline/

[19] https://cnbc.com/2020/10/06/what-the-workforce-will-look-like-in-2025-as-it-morphs-due-to-pandemic.html

[20] Burger B, Maffettone P M and Gusev V V *et al* 2020 A mobile robotic chemist *Nature* **583** 237–41

[21] https://capgemini.com/wp-content/uploads/2021/03/The-Future-of-Work_Final.pdf

[22] Cmswire.com/digital-workplace/the-elements-of-sustainable-collaboration

[23] Fayard A-L, Weeks J and Khan M 2021 Designing the hybrid office *Harv. Bus. Rev.* (March–April)

Milton Keynes UK
Ingram Content Group UK Ltd.
UKHW050223050324
438897UK00003B/11